CAMBRIDGE LIBRARY COLLECTION

Books of enduring scholarly value

Physical Sciences

From ancient times, humans have tried to understand the workings of the world around them. The roots of modern physical science go back to the very earliest mechanical devices such as levers and rollers, the mixing of paints and dyes, and the importance of the heavenly bodies in early religious observance and navigation. The physical sciences as we know them today began to emerge as independent academic subjects during the early modern period, in the work of Newton and other 'natural philosophers', and numerous sub-disciplines developed during the centuries that followed. This part of the Cambridge Library Collection is devoted to landmark publications in this area which will be of interest to historians of science concerned with individual scientists, particular discoveries, and advances in scientific method, or with the establishment and development of scientific institutions around the world.

Sound

Professor of natural philosophy at the Royal Institution between 1853 and 1887, the British physicist and mountaineer John Tyndall (1820–93) passionately sought to share scientific understanding with the Victorian public. A lucid and highly regarded communicator, he lectured on such topics as heat, light, magnetism and electricity. In this collection of eight lectures, first published in 1867, Tyndall explains numerous acoustic phenomena for a non-specialist audience. Emphasising the practical nature of scientific enquiry, he describes experiments throughout and includes many illustrations of laboratory equipment. The lectures discuss the general properties of sound, how it travels, how noise and music differ, how gas flames can produce musical notes, and much more. Several of Tyndall's other publications, from his work on radiant heat to his exploration of alpine glaciers, are also reissued in this series.

Cambridge University Press has long been a pioneer in the reissuing of out-of-print titles from its own backlist, producing digital reprints of books that are still sought after by scholars and students but could not be reprinted economically using traditional technology. The Cambridge Library Collection extends this activity to a wider range of books which are still of importance to researchers and professionals, either for the source material they contain, or as landmarks in the history of their academic discipline.

Drawing from the world-renowned collections in the Cambridge University Library and other partner libraries, and guided by the advice of experts in each subject area, Cambridge University Press is using state-of-the-art scanning machines in its own Printing House to capture the content of each book selected for inclusion. The files are processed to give a consistently clear, crisp image, and the books finished to the high quality standard for which the Press is recognised around the world. The latest print-on-demand technology ensures that the books will remain available indefinitely, and that orders for single or multiple copies can quickly be supplied.

The Cambridge Library Collection brings back to life books of enduring scholarly value (including out-of-copyright works originally issued by other publishers) across a wide range of disciplines in the humanities and social sciences and in science and technology.

Sound

A Course of Eight Lectures
Delivered at
the Royal Institution of Great Britain

JOHN TYNDALL

CAMBRIDGE
UNIVERSITY PRESS

University Printing House, Cambridge, CB2 8BS, United Kingdom

Published in the United States of America by Cambridge University Press, New York

Cambridge University Press is part of the University of Cambridge.
It furthers the University's mission by disseminating knowledge in the pursuit of education, learning and research at the highest international levels of excellence.

www.cambridge.org
Information on this title: www.cambridge.org/9781108067386

This edition first published 1867
This digitally printed version 2014

ISBN 978-1-108-06738-6 Paperback

ON

SOUND.

LONDON
PRINTED BY SPOTTISWOODE AND CO.
NEW-STREET SQUARE

H. Adlard. Sc

Chladni.

London: Longmans & Co.

SOUND.

A COURSE OF

EIGHT LECTURES

DELIVERED AT

THE ROYAL INSTITUTION OF GREAT BRITAIN

BY

JOHN TYNDALL, LL.D. F.R.S.

PROFESSOR OF NATURAL PHILOSOPHY IN THE
ROYAL INSTITUTION AND IN THE ROYAL
SCHOOL OF MINES.

LONDON:
LONGMANS, GREEN, AND CO.
1867.

To the Memory of

MY FRIEND

RICHARD DAWES

LATE DEAN OF HEREFORD

This Book is Dedicated.

1867. J. T.

PREFACE.

IN the following pages I have tried to render the science of Acoustics interesting to all intelligent persons, including those who do not possess any special scientific culture.

The subject is treated experimentally throughout, and I have endeavoured so to place each experiment before the reader, that he should realise it as an actual operation. My desire indeed has been to give distinct images of the various phenomena of acoustics, and to cause them to be seen mentally in their true relations.

I have been indebted to the kindness of some of my English friends for a more or less complete examination of the proof sheets of this work. To one celebrated German friend, who has given himself the trouble of reading the proofs from beginning to end, my special thanks are due and tendered.

There is a growing desire for scientific culture throughout the civilised world. The feeling is natural, and, under the circumstances, inevitable. For a power which influences so mightily the intellectual and material action of the age, could not fail to arrest attention and challenge examination. In our schools and universities a movement in favour of science has begun which, no doubt, will end in the recognition of its claims, both as a source of knowledge and as a means of discipline. If by showing, however inadequately, the features and the mien of physical science to men of influence who derive their culture from another source, this book should indirectly aid those engaged in the movement referred to, it will not have been written in vain.

I have placed in front of the book a portrait of Chladni, for which, and the autograph underneath it, I am indebted to my eminent friend Professor W. Weber of Gottingen. It was gratifying to me to find my estimate of Chladni confirmed by Professor Weber, who thus writes to me regarding the great acoustician: 'I knew Chladni personally. From my youth up, he was my leader and model as a man of science, and I cannot too thankfully acknowledge the influence which his stimulating encouragement, during the last years of his life, had upon my own scientific labours.'

Four years ago a work was published by Professor Helmholtz, entitled 'Die Lehre von den Tonempfindungen, to the scientific portion of which I have given considerable attention. Copious references to it will be found in the following pages; but they fail to give an adequate idea of the thoroughness and excellence of the work. To those especially who wish to pursue the subject into its æsthetic developments, the Third Part of the Tonempfindungen cannot fail to be of the highest interest and use.

Finally, I have ventured to connect this book with the name of a man, who, had he lived, would have been the first to turn it to good account; who blended in his own beautiful character the wisdom of mature years with the spring-like freshness of a boy. When together indeed, we were men and boys by turns. This union of life, love, and wisdom rendered Richard Dawes a great educator of the young, in which capacity, and to the incalculable profit of the village children on whom his influence fell, he nobly and beneficently spent his life.

The Illustrations of this work were for the most part drawn for me by Mr. Becker, to whose ability as a mechanician and to whose skill as a draughtsman I am continually indebted. The wood engravings were executed by Mr. Branston, and the portrait of Chladni by Mr. H. Adlard.

CONTENTS.

LECTURE I.

LECTURE II.

LECTURE III.

LECTURE IV.

LECTURE V.

LECTURE VI.

LECTURE VII.

LECTURE VIII.

SOUND.

LECTURE I.

THE NERVES AND SENSATION—PRODUCTION AND PROPAGATION OF SONOROUS MOTION—EXPERIMENTS ON SOUNDING BODIES PLACED IN VACUO—ACTION OF HYDROGEN ON THE VOICE—PROPAGATION OF SOUND THROUGH AIR OF VARYING DENSITY — REFLECTION OF SOUND — ECHOES — REFRACTION OF SOUND—INFLECTION OF SOUND; CASE OF ERITH VILLAGE AND CHURCH —INFLUENCE OF TEMPERATURE ON VELOCITY — INFLUENCE OF DENSITY AND ELASTICITY—NEWTON'S CALCULATION OF VELOCITY—THERMAL CHANGES PRODUCED BY THE SONOROUS WAVE—LAPLACE'S CORRECTION OF NEWTON'S FORMULA—RATIO OF SPECIFIC HEATS AT CONSTANT PRESSURE AND AT CONSTANT VOLUME DEDUCED FROM VELOCITIES OF SOUND—MECHANICAL EQUIVALENT OF HEAT DEDUCED FROM THIS RATIO—INFERENCE THAT ATMOSPHERIC AIR POSSESSES NO SENSIBLE POWER TO RADIATE HEAT—VELOCITY OF SOUND IN DIFFERENT GASES — VELOCITY IN LIQUIDS AND SOLIDS—INFLUENCE OF MOLECULAR STRUCTURE ON THE VELOCITY OF SOUND.

THE various nerves of the human body have their origin in the brain, and the brain is the seat of sensation. When you wound your finger, the nerves which run from the finger to the brain convey intelligence of the injury, and if these nerves be severed, however serious the hurt may be, no pain is experienced. We have the strongest reason for believing that what the nerves convey to the brain is in all cases *motion*. It is the motion excited by sugar in the nerves of taste which, transmitted to the brain, produces the sensation of sweetness, while bitterness is the result of the motion produced by aloes. It is the motion excited in the olfactory nerves by the effluvium of a rose, which announces itself in the brain as the odour of the rose. It is the motion imparted by the sunbeams to the optic nerve which, when it reaches the brain, awakes the consciousness of light; while a similar motion imparted to other nerves resolves itself into heat in the same wonderful organ.*

The motion here meant is not that of the nerve as a whole; it is the vibration, or tremor, of its molecules or smallest particles.

Different nerves are appropriated to the transmission of different kinds of molecular motion. The nerves of taste, for example, are not competent to transmit the tremors of light, nor is the optic nerve competent to transmit sonorous vibration. For this latter a special nerve is necessary, which passes from the brain into one of the cavities of the ear, and there spreads out in a multitude of filaments. It is the motion imparted to this, the *auditory nerve*, which, in the brain, is translated into sound.

We have this day to examine how sonorous motion is produced and propagated. Applying a flame to this small collodion balloon, which contains a mixture of oxygen and hydrogen, the gases explode, and every ear in this room is conscious of a shock, to which the name of sound is given. How was this shock transmitted from the balloon to your organs of hearing? Have our exploding gases shot the air-particles against the auditory nerve as a gun shoots a ball against a target? No doubt, in the neighbourhood of the balloon, there is to some extent a propulsion of particles; but air shooting through air comes speedily to rest, and no particle of air from the vicinity of the balloon reached the ear of any person here present. The

* The rapidity with which an impression is transmitted through the nerves, as first determined by Helmholtz and confirmed by Du Bois Raymond, is 93 feet a second.

process was this:—When the flame touched the mixed gases they combined chemically, and their union was accompanied by the development of intense heat. The air at this hot focus expanded suddenly, forcing the surrounding air violently away on all sides. This motion of the air close to the balloon was rapidly imparted to that a little further off, the air first set in motion coming at the same time to rest. The air, at a little distance, passed its motion on to the air at a greater distance, and came also in its turn to rest. Thus each shell of air, if I may use the term, surrounding the balloon, took up the motion of the shell next preceding, and transmitted it to the next succeeding shell, the motion being thus propagated as a *pulse* or *wave* through the air.

In air at the freezing temperature this pulse is propagated with a speed of 1,090 feet a second.

The motion of the pulse must not be confounded with the motion of the particles which at any moment constitute the pulse. For while the wave moves forward through considerable distances, each particular particle of air makes only a small excursion to and fro.

Fig. 1.

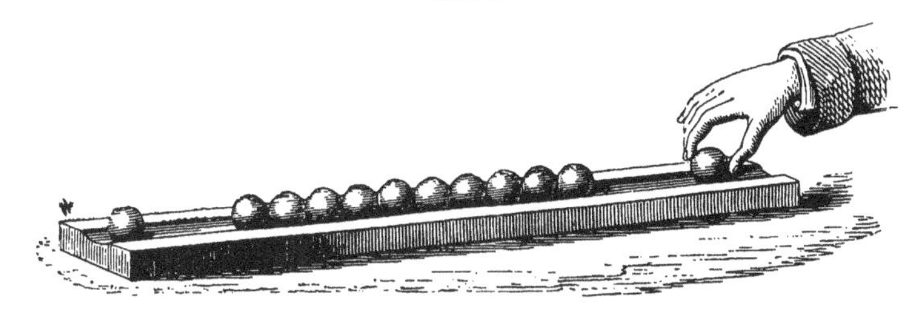

The process may be rudely represented by the propagation of motion through a row of glass balls, such as are employed in the game of *solitaire*. I place these balls along a groove thus, fig. 1, each of them touching its neighbours. Taking one of them in my hand, I urge it against the end of the row. The motion thus imparted to the first ball is delivered up to the second, the motion of the second is delivered up to the third, the motion of the third is imparted to the fourth; each ball, after having given up its motion, returning itself to rest. The last ball only of the row flies away. Thus is sound conveyed from particle to particle through the air. The particles which fill the cavity of the ear are finally driven against *the tympanic membrane,* which is stretched across the passage leading to the brain. This membrane, which closes the 'drum' of the ear, is thrown into vibration, its motion is transmitted to the ends of the auditory nerve, and afterwards along the nerve to the brain, where the vibrations are translated into sound. How it is that the motion of the nervous matter can thus excite the consciousness of sound is a mystery which we cannot fathom.

Let me endeavour to illustrate the propagation of sound by another homely but useful illustration. I have here

Fig. 2.

five young assistants, A, B, C, D, and E, fig. 2, placed in a row, one behind the other, each boy's hands resting against the back of the boy in front of him. E is now foremost, and A finishes the row behind. I suddenly push A; A pushes

B, and regains his upright position; B pushes C; C pushes D; D pushes E; each boy, after the transmission of the push, becoming himself erect. E, having nobody in front, is thrown forward. Had he been standing on the edge of a precipice, he would have fallen over; had he stood in contact with a window, he would have broken the glass; had he been close to a drum-head, he would have shaken the drum. We could thus transmit a push through a row of a hundred boys, each particular boy, however, only swaying to and fro. Thus, also, we send sound through the air, and shake the drum of a distant ear, while each particular particle of the air concerned in the transmission of the pulse makes only a small oscillation.

Scientific education ought to teach us to see the invisible as well as the visible in nature; to picture with the eye of the mind those operations which entirely elude the eye of the body; to look at the very atoms of matter in motion and at rest, and to follow them forth, without ever once losing sight of them, into the world of the senses, and see them there integrating themselves in natural phenomena. With regard to the point now under consideration, you will, I trust, endeavour to form a definite image of a wave of sound. You ought to see mentally the air particles when urged outwards by the explosion of our balloon crowding closely together; but immediately behind this condensation you ought to see the particles separated more widely apart. You ought, in short, to be able to seize the conception that a sonorous wave consists of two portions, in the one of which the air is more dense, and in the other of which it is less dense than usual. A condensation and a rarefaction, then, are the two constituents of a wave of sound.*

Let us turn once more to our row of boys, for we have

* A sonorous wave will be more strictly defined in Lecture II.

not yet extracted from them all that they can teach us. When I push A, he may yield languidly, and thus tardily deliver up the motion to his neighbour B. B may do the same to C, C to D, and D to E. In this way the motion may be transmitted with comparative slowness along the line. But A, when I push him, may, by a sharp muscular effort and sudden recoil, deliver up promptly his motion to B, and come himself to rest; B may do the same to C, C to D, and D to E, the motion being thus transmitted rapidly along the line. Now this sharp muscular effort and sudden recoil is analogous to the *elasticity* of the air in the case of sound. In a wave of sound, a lamina of air, when urged against its neighbour lamina, delivers up its motion and recoils, in virtue of the elastic force exerted between them; and the more rapid this delivery and recoil, or in other words the greater the elasticity of the air, the greater is the velocity of the sound.

But if air be thus necessary to the propagation of sound, what must occur when a sonorous body, a ringing bell for example, is placed in a space perfectly void of air? Out of that space the sound could never come. The hammer might strike, but it would strike silently. A celebrated experiment which proved this was made by a philosopher named Hawksbee, before the Royal Society in 1705.* He so fixed a bell within the receiver of an air-pump, that he could ring the bell when the receiver was exhausted. Before the air was withdrawn the sound of the bell was heard within the receiver; after the air was withdrawn the sound became so faint as to be hardly perceptible. I have here an arrangement which enables me to repeat, in a very perfect manner, the experiment of Hawksbee. Within this jar, G G′, fig. 3, resting on the plate of an air-pump is a bell, B, associated with

* Philosophical Transactions, 1705.

clockwork.* I exhaust the jar as perfectly as possible, and now, by means of a rod, *r r′*, which passes air-tight through the top of the vessel, I loose the detent which holds the hammer. It strikes, and you see it striking, but only those close to the bell can hear the sound. I now allow hydrogen gas, which you know is fourteen times lighter than air, to enter the vessel. The sound of the bell is not sensibly augmented by the presence of this attenuated gas, though the receiver is now full of it.† By working the pump, the atmosphere round the bell is rendered still more attenuated. In this way we obtain a vacuum more perfect than that of Hawksbee, and this is important, for it is the last traces of air that are chiefly effective in this experiment. You now see the hammer pounding the bell, but you hear no sound. Even when I place my ear against the exhausted receiver, I am unable to hear the faintest tinkle. Observe also that the bell is suspended by strings, for if it were allowed to rest upon the plate of the air-pump, the vibrations would communicate themselves to the plate, and be transmitted

Fig. 3.

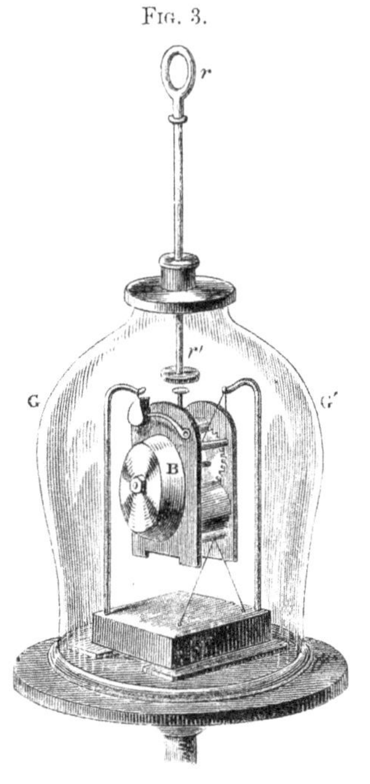

* A very effective instrument presented to the Royal Institution by Mr. Warren De la Rue.

† Leslie, I believe, was the first to notice this. Air, it may be stated, reduced to the specific gravity of hydrogen, transmits the sound of the bell vastly better than hydrogen. A whole atmosphere of this gas has no sensible effect in restoring the sound of the bell, while the fifteenth of an atmosphere of air renders its ringing very audible.

to the air outside. All that I can hear by the most concentrated attention, with my ear placed against the receiver, is a feeble thud, due to the transmission of the shock of the hammer through the strings which support the bell. I now permit air to enter the jar with as little noise as possible. You immediately hear a feeble sound, which grows louder as the air becomes more dense; and now every person in this large assembly distinctly hears the ringing of the bell.*

At great elevations in the atmosphere sound is sensibly diminished in loudness. De Saussure thought the explosion of a pistol at the summit of Mont Blanc to be about equal to that of a common cracker below. I have several times repeated this experiment: first, in default of anything better, with a little tin cannon, the torn remnants of which are now before you, and afterwards with pistols. What struck me was the absence of that density and sharpness in the sound which characterise it at lower elevations. The pistol-shot resembled the explosion of a champagne bottle, but it was still loud. The withdrawal of half an atmosphere does not very materially affect our ringing bell, and air of the density found at the top of Mont Blanc is still capable of powerfully affecting the auditory nerve. That highly attenuated air is able to convey sound of great intensity, is forcibly illustrated by the explosion of meteorites at great elevations above the earth. Here, however, the initial disturbance must be exceedingly violent.

The motion of sound, like all other motion, is enfeebled by its transference from a light body to a heavy one. I remove the receiver which has hitherto covered our bell;

* By directing the beam of an electric lamp on glass bulbs filled with a mixture of equal volumes of chlorine and hydrogen, I have caused the bulbs to explode in vacuo and in air. The difference, though not so striking as I at first expected, was perfectly distinct.

you hear how much more loudly it rings in the open air. When the bell was covered the aerial vibrations were first communicated to the heavy glass jar, and afterwards by the jar to the air outside; a great diminution of intensity being the consequence. The action of hydrogen gas upon the voice is an illustration of the same kind. The voice is formed by urging air from the lungs through an organ called the larynx. In its passage it is thrown into vibration by the vocal chords which thus generate sound. But when I fill my lungs with hydrogen, and endeavour to speak, the vocal chords impart their motion to the hydrogen, which transfers it to the outer air. By this transference from a light gas to a heavy one, the sound is weakened in a remarkable degree.* The consequence is very curious. You have already formed a notion of the strength and quality of my voice. I now empty my lungs of air, and inflate them with hydrogen from this gasholder. I try to speak vigorously, but my voice has lost wonderfully in power, and changed wonderfully in quality. You hear it, hollow, harsh, and unearthly: I cannot otherwise describe it.

The intensity of a sound depends on the density of the air in which the sound is generated, and not on that of the air in which it is heard.† Supposing the summit of Mont Blanc to be equally distant from the top of the Aiguille Verte and the bridge at Chamouni; and supposing two observers stationed, the one upon the bridge and the other upon the Aiguille: the sound of a cannon fired on Mont Blanc would reach both observers with the same intensity, though in the one case the sound would pursue its way through the rare air above, while in the other it would descend through the denser air below. Again, let a

* It may be that the gas fails to throw the vocal chords into sufficiently strong vibration. The *laryngoscope* might decide this question.

† Poisson Mécanique, vol. ii. p. 707.

straight line equal to that from the bridge of Chamouni to the summit of Mont Blanc, be measured along the earth's surface in the valley of Chamouni, and let two observers be stationed, the one on the summit and the other at the end of the line; the sound of a cannon fired on the bridge would reach both observers with the same intensity, though in the one case the sound would be propagated through the dense air of the valley, and in the other case would ascend through the rarer air of the mountain. Charge two cannon equally, and fire one of them at Chamouni, and the other at the top of Mont Blanc, the one fired in the heavy air below may be heard above, while the one fired in the light air above is unheard below; because, at its origin, the sound generated in the denser air is louder than that generated in the rarer.

You have, I doubt not, a clear mental picture of the propagation of the sound from our exploding balloon through the surrounding air. The wave of sound expands on all sides, the motion produced by the explosion being thus diffused over a continually augmenting mass of air. It is perfectly manifest that this cannot occur without an enfeeblement of the motion. Take the case of a shell of air of a certain thickness, with a radius of one foot, reckoned from the centre of explosion. A shell of air of the same thickness, but of two feet radius, will contain four times the quantity of matter; if its radius be three feet, it will contain nine times the quantity of matter; if four feet, it will contain sixteen times the quantity of matter, and so on. Thus the quantity of matter set in motion augments as the square of the distance from the centre of explosion. The *intensity* or loudness of the sound diminishes in the same proportion. We express this law by saying that the intensity of the sound varies inversely as the square of the distance.

Let us look at the matter in another light. The mechanical effect of a ball striking a target depends on two things, the weight of the ball, and the velocity with which it moves. The effect is proportional to the weight simply; but it is proportional to the square of the velocity. The proof of this is easy, but it belongs to ordinary mechanics rather than to our present subject. Now what is true of the cannon-ball striking a target, is also true of an air-particle striking the tympanic membrane of the ear. Fix your attention upon a particle of air as the sound-wave passes over it; it is urged from its position of rest towards a neighbour particle, first with an accelerated motion, and then with a retarded one. The force which first urges it, is opposed by the elastic force of the air, which finally stops the particle, and causes it to recoil. At a certain point of its excursion, the velocity of the particle is a maximum. *The intensity of the sound is proportional to the square of this maximum velocity.* The sonorous effect is expressed by the same law as the mechanical effect. All, in fact, that goes on outside of ourselves is reducible to pure mechanics, and if we hear one sound louder than another, it is because our nerves are hit harder in the one case than in the other.

The distance through which the air-particle moves to and fro, when the sound-wave passes it, is called the *amplitude* of the vibration. The intensity of the sound is also proportional to the square of the amplitude.

This weakening of the sound, according to the law of inverse squares, would not take place if the sound-wave were so confined as to prevent its lateral diffusion. By sending it through a tube with a smooth interior surface we accomplish this, and the wave thus confined may be transmitted to great distances with very little diminution of intensity. Into one end of this tin tube, fifteen feet long, I whisper in a manner quite inaudible to the people nearest to me, but a listener at the other end hears me distinctly. I place my watch at one end of the tube; the person at the other end hears the ticks, though nobody else does. At the distant end of the tube I now place a lighted candle, *c*, fig. 4. When I clap my hands at this end, the flame instantly ducks down. It is not quite extinguished, but it is forcibly depressed. When I clap two books, B B′, together, I blow the candle out.* You may here observe, in a rough way, the speed with which the sound-wave is propagated. The instant I clap, the flame is extinguished; there is no sensible interval between the clap and the extinction of the

FIG. 4.

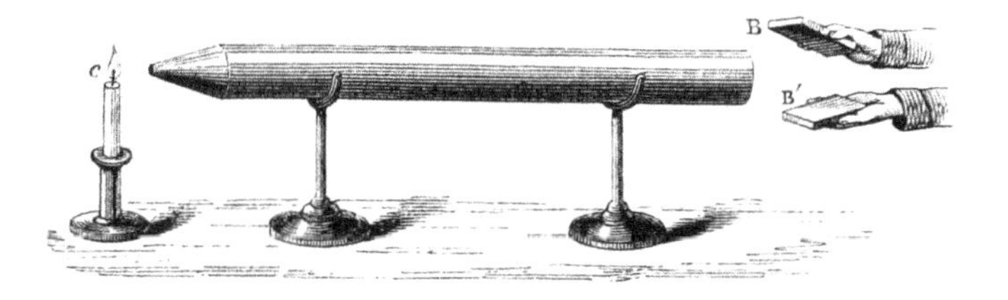

flame. I do not say that the time required by the sound to travel through this tube is immeasurably short, but simply that the interval is too short for your senses to appreciate it. To show you that it is a *pulse* and not a *puff* of air, I fill one end of the tube with the smoke of brown paper. On clapping the books together no trace of this smoke is ejected from the other end. The pulse has passed through both smoke and air without carrying either of them along with it.

Further, I hold in my hand the open end of a gutta-percha tube of about an inch interior diameter. It passes through the floor to a room underneath this one, and

* To converge the pulse upon the flame, the tube was caused to end in a cone.

thence through a window into the yard of the Royal Institution, its end there being opened out into a funnel. I send an assistant down to the yard. He may shout there, but as long as he keeps away from the open end of this tube we cannot hear him. I however hear his lightest whisper when he speaks into the tube. I address him through the tube, asking him, 'Are you ready?' You hear his reply. I desire him to pitch his voice so as to produce a continuous musical note, and here is the music issuing in the midst of us. By alternately closing and opening the tube I produce this babble, for the mere placing of my thumb against the open end of the tube cuts off the sound. The man now blows a horn, directing its sound into the tube, and here you have its notes poured forth. I now call on him to sing, and you hear him sing the national anthem, the sound of his voice appearing to come not from a distance, but from the throat of a little performer hidden immediately within this end of the tube.

The celebrated French philosopher, Biot, observed the transmission of sound through the empty waterpipes of Paris, and found that he could hold a conversation in a low voice through an iron tube 3120 feet in length. The lowest possible whisper, indeed, could be heard at this distance, while the firing of a pistol into one end of the tube quenched a lighted candle at the other.

The action of sound thus illustrated is exactly the same as that of light and radiant heat. They, like sound, are wave motions. Like sound they diffuse themselves in space, diminishing in intensity according to the same law. Like sound also, light and radiant heat, when sent through a tube with a reflecting interior surface, may be conveyed to great distances with comparatively little loss. In fact, every experiment on the reflection of light, has its analogue in the reflection of sound. Permit me to illustrate this analogy by one or two additional experiments. On yonder gallery you see an electric lamp, placed close to the clock of this lecture room. An assistant in the gallery ignites

FIG. 5.

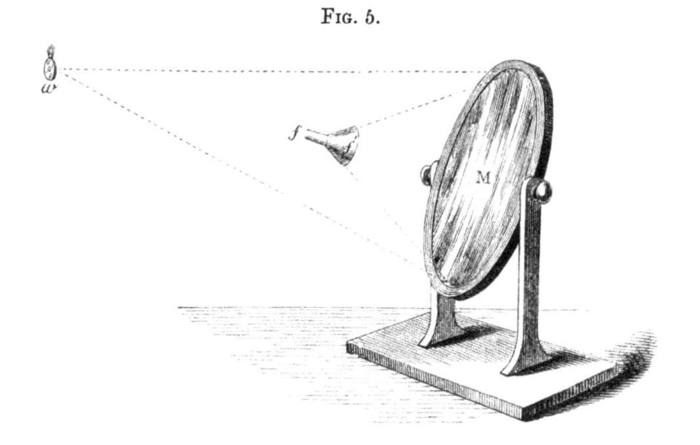

the lamp, and directs its powerful beam upon a mirror, M, fig. 5, placed here behind the lecture table. By the act of reflection the divergent beam is converted into this

FIG. 6.

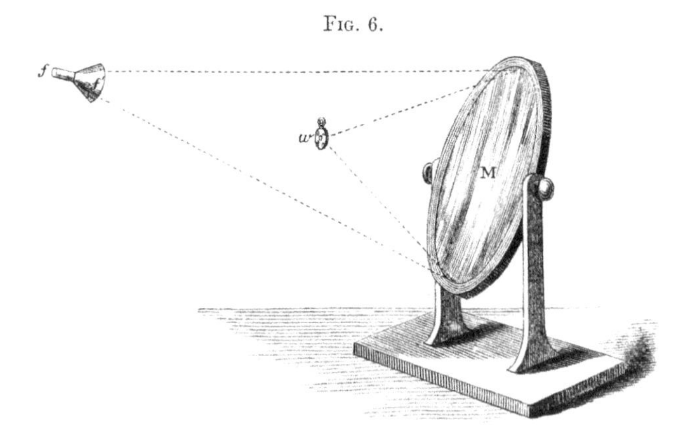

splendid luminous cone. I mark the point of convergence; and the lamp being extinguished, I place my ear at that point. Here every sound-wave sent forth by the clock, and reflected by the mirror, is gathered up, and I now hear the ticks as if they came, not from the clock, but from the mirror. I will stop the clock, and have a watch, *w*, fig. 5, held at the place occupied a moment ago by the electric light. At this great distance I distinctly hear the ticking of the watch. My hearing is much aided by introducing the end *f* of a glass funnel into my ear, the funnel here acting the part of an ear-trumpet. We know, moreover, that in optics the positions of a body and of its image are reversible. I place a candle at this lower focus; you see its image on the gallery above, and I have only to turn the mirror on its stand, to make the image fall upon any one of the row of persons who occupy the front seat in the gallery. Removing the candle, and putting the watch, *w*, fig. 6, in its place, the person on whom the image of the candle fell, distinctly hears the ticking. When the ear is assisted by the glass funnel, the reflected ticks of the clock in our first experiment are so powerful as to suggest the idea of something pounding against the tympanum, while the direct ticks are scarcely, if at all, heard.

Here, finally, are two parabolic mirrors, one of them, $n\,n'$, fig. 7, placed upon the table, and the other, $m\,m'$, drawn up to the ceiling of this theatre; they are five-and-twenty feet apart. I first place the carbon points of the electric light in the focus *a*, of the lower mirror, and ignite them. A fine luminous cylinder rises like a pillar to the upper mirror, which brings the parallel beam to a focus. You see at that focus a spot of sunlike brilliancy, due to the reflection of the light, from the surface of a watch, *w*, there suspended. I remove the electric light, the watch is ticking, though in my present position I do not hear it. At this lower focus, *a*, however, we have the energy of every sonorous wave converged. Placing the ear at *a*, the ticking is as audible as if the watch were at hand; the sound, as in the former case, appearing to proceed, not from the watch itself, but from the lower mirror.*

Curved roofs and ceilings act as mirrors upon sound. In our laboratory, for example, the singing of a kettle seems, in certain positions, to come, not from the fire on which it is placed, but from the ceiling. Inconvenient secrets have been thus revealed, an instance of which has been cited by Sir John Herschel.† In one of the cathedrals in Sicily, the confessional was so placed that the whispers of the penitents were reflected by the curved roof, and brought to a focus at a distant part of the edifice. The focus was discovered by accident, and for some time the person who discovered it took pleasure in hearing, and in bringing his friends to hear, utterances intended for the priest alone. One day, it is said, his own wife occupied the penitential stool, and both he and his

FIG. 7.

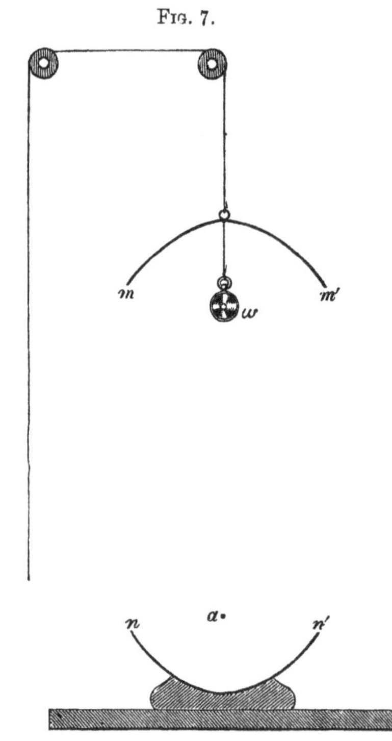

* It is recorded that a bell placed on an eminence in Heligoland failed, on account of its distance, to be heard in the town. A parabolic reflector placed behind the bell so as to reflect the sound-waves in the direction of the long sloping street, caused the strokes of the bell to be distinctly heard at all times.

† Ency. Met. art. 'Sound.'

friends were thus made acquainted with secrets which were the reverse of amusing to one of the party.

When a sufficient interval exists between a direct and a reflected sound, we hear the latter as an *echo.* The reflected sound moves with the same velocity as the direct sound, so that in air of the freezing temperature the echo of a pistol-shot from the face of a cliff 1,090 feet distant is heard two seconds after the explosion.

But sound, like light, may be reflected several times in succession, and as the reflected light under these circumstances becomes gradually feebler to the eye, so the successive echoes become gradually feebler to the ear. In mountain regions this repetition and decay of sound produces wonderful and pleasing effects. Visitors to Killarney will remember the fine echo in the Gap of Dunloe. When a trumpet is sounded at the proper place in the Gap, the sonorous waves reach the ear in succession after one, two, three, or more reflections from the adjacent cliffs, and thus die away in the sweetest cadences. There is a deep *cul-de-sac,* called the Ochsenthal, formed by the great cliffs of the Engelhörner, near Rosenlaui, in Switzerland, where the echoes warble in a wonderful manner. The sound of the Alpine horn also rebounding from the rocks of the Wetterhorn or the Jungfrau, is in the first instance heard roughly. But by successive reflections, the notes are rendered more soft and flute-like, the gradual diminution of intensity giving the impression that the source of sound is retreating further and further into the solitudes of ice and snow.

In large unfurnished rooms the reflection of sound sometimes produces very curious effects. Standing, for example, in the gallery of the Bourse at Paris, you hear the confused vociferation of the excited multitude below. You see all the motions—of their lips as well as of their hands and arms. You know they are speaking—often indeed with vehemence, but what they say you know not.

The voices mix with their echoes into a chaos of noise, out of which no intelligible utterance can emerge. The echoes of a room are materially damped by its furniture. The presence of an audience may also render intelligible speech possible, where, without an audience, the definition of the direct voice is destroyed by its echoes. On the 16th of May, 1865, having to lecture in the Senate House of the University of Cambridge, I first made some experiments as to the loudness of voice necessary to fill the room, and was dismayed to find that a friend placed at a distant part of the hall could not follow me because of the echoes. The assembled audience, however, so quenched the sonorous waves, that the echoes were practically absent, and my voice was plainly heard in all parts of the Senate House.

Sounds are also reflected from the clouds. When the sky is clear, the report of a cannon on an open plain is short and sharp; while a cloud is sufficient to produce an echo like the rolling of distant thunder. A feeble echo also occurs when sound passes from one mass of air to another of different density. Humboldt relates that from a certain position on the plains of Antures, the sound of the great falls of the Orinoco resembles the beating of a surf upon a rocky shore, being much louder by night than by day. This is not due to the greater stillness of the night, for the hum of insects and the roar of beasts rendered the night much noisier than the day. Humboldt thus explained the observation:—Between him and the falls lay a vast grassy plain, with multitudes of bare rocks protruding from it. When exposed to the sun, these rocks assumed a temperature far higher than that of the adjacent grass; over each of them therefore rose a column of heated air, less dense than that which surrounded it. Thus by day the sound had to pass through an atmosphere which frequently changed its density; the partial echoes at the limiting surfaces of rare and dense air were incessant, and the sound was consequently enfeebled. At night those differences of temperature ceased to exist, and the sound-waves, travelling through a homogeneous atmosphere, reached the ear undiminished by reflection. The case has its parallel in light, which is also reflected at the limiting surfaces of different optical media; so that a mixture of different media, each of itself transparent, may intercept light. Thus by the mixture of air and water foam is produced, which in moderate thicknesses is impervious to light, in consequence of repeated reflection. All colourless transparent substances, when reduced to powder, are white and opaque for the same reason.

Thunder-peals are unable to penetrate the air to a distance commensurate with their intensity, because of the non-homogeneous character of the atmosphere which accompanies thunder-storms. From the same cause battles have raged, and been lost, within a short distance of the reserves of the defeated army, while they were waiting for the sound of artillery to call them to the scene of action. Falling snow has been often referred to as offering a great hindrance to the passage of sound, but I imagine it to be less obstructive than is usually supposed. Sound appears to make its way freely between the falling flakes. On the 29th of December, 1859, I traced a line across the Mer de Glace of Chamouni, at an elevation of nearly seven thousand feet above the sea. The glacier there is half a mile wide, and during the setting out of the line snow fell heavily. I have never seen the atmosphere in England so thickly laden. Still I was able to see through the storm quite across the glacier, and also to make my voice heard. When close to the opposite side, one of the assistants chanced to impede my view. I called out to him to stand aside, and he did so immediately. At the end of the line the men shouted '*nous sommes finis,*' and I distinctly heard them through the half-mile of falling snow.

Sir John Herschel, in his excellent article 'Sound' in the *Encyclopædia Metropolitana,* has collected with others the following instances of echoes:—An echo in Woodstock Park repeats seventeen syllables by day, and twenty by night; one on the banks of the Lago del Lupo, above the fall of Terni, repeats fifteen. The tick of a watch may be heard from one end of the abbey church of St. Albans to the other. In Gloucester Cathedral, a gallery of an octagonal form conveys a whisper seventy-five feet across the nave. In the whispering gallery of St. Paul's, the faintest sound is conveyed from one side to the other of the dome, but is not heard at any intermediate point. At Carisbrook Castle, in the Isle of Wight, is a well 210 feet deep, and 12 wide. The interior is lined by smooth masonry; when a pin is dropped into the well, it is distinctly heard to strike the water. I may add that shouting or coughing into this well produces a resonant ring of some duration.*

I have now to point out another important analogy between sound and light, which has been established by M. Sondhauss.† I ignite our electric lamp, and place in front of it this fine large lens; the lens compels the rays of light that fall upon it to deviate from their direct and divergent course, and to form this convergent cone behind it. This *refraction* of the luminous beam is a consequence of the retardation suffered by the light in passing through the glass. Sound may be similarly refracted by causing it to pass through a lens which retards its motion. Such a

* Placing himself close to the upper part of the wall of the London Colosseum, a circular building 130 feet in diameter, Mr. Wheatstone found a word pronounced to be repeated a great many times. A single exclamation appeared like a peal of laughter, while the tearing of a piece of paper was like the patter of hail.

† *Poggendorff's Annalen,* vol. lxxxv. p. 378; *Philosoph. Mag.* vol. v p. 73.

lens is formed when we fill a thin balloon with some gas heavier than air. Here, for example, is a collodion balloon B, fig. 8, filled with carbonic acid gas, the envelope being so thin as to yield readily to the pulses which strike against

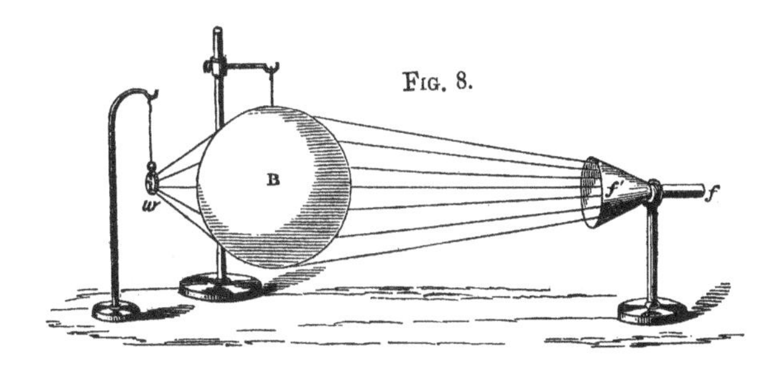

Fig. 8.

it, transmitting them to the gas inside.* I now hang up my watch, *w*, close to the lens, beyond which, and at a distance of four or five feet from the lens, I place my ear assisted by the glass funnel *f f′*. By moving my head about, I soon discover a position in which the ticking is particularly loud. This, in fact, is the focus of the lens. If I move my ear from this focus the intensity of the sound falls; if when my ear is at the focus the balloon be removed, the ticks are enfeebled; on replacing the balloon their force is restored. The lens, in fact, enables me to hear the ticks distinctly when they are perfectly inaudible to the unaided ear.

Fig. 9.

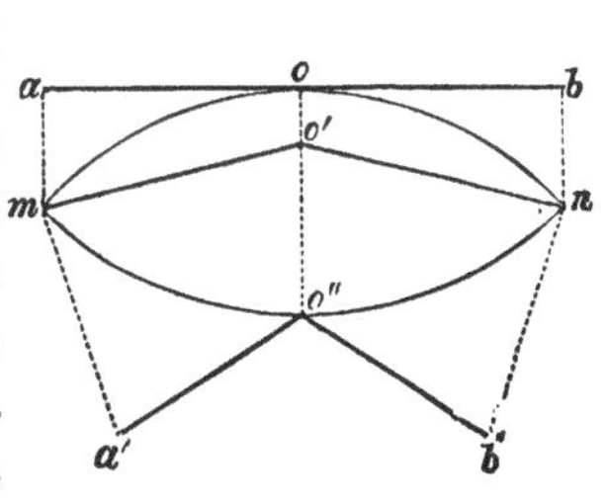

How a sound-wave is thus converged may be comprehended by reference to fig. 9. Let *m n* be a section of the sound-lens, and *a b* a portion of a sonorous wave approaching

* Thin india-rubber balloons also form excellent sound-lenses.

it from a distance. The middle point, o, of the wave first touches the lens, and is first retarded by it. By the time the ends a and b, still moving through air, reach the balloon, the middle point o, pursuing its way through the heavier gas within, will have only reached o'. The wave is therefore broken at o', and the direction of motion being at right angles to the face of the wave, the two halves will now encroach upon each other. The convergence of the two halves of the wave is augmented on quitting the lens. For when o' has reached o'', the two ends a and b will have pushed forward to a greater distance, say to a' and b'. Soon afterwards the two halves of the wave will cross each other, or, in other words, come to a focus, the air at the focus being agitated by the sum of the motions of the two waves.*

When a long sea-roller meets an isolated rock in its passage, it rises against the rock, and embraces it all round. Facts of this nature caused Newton to reject the undulatory theory of light. He contended that if light were a product of wave motion we could have no shadows, because the waves of light would propagate themselves round opaque bodies as a wave of water round a rock. It has been proved since his time that the waves of light do bend round opaque bodies; but with that we have nothing now to do. A sound-wave certainly bends thus round an obstacle, though as it diffuses itself in the air at the back of the obstacle it is enfeebled in power, the obstacle thus producing a partial *shadow* of the sound. Anybody who has ever heard a railway train approach through cuttings and along embankments, will have noticed great variations in the intensity of the sound. The interposition of a hill in the Alps suffices to diminish materially

* For the sake of simplicity, I have shown the wave broken at o and its two halves straight. The surface of the wave, however, is really a curve, with its concavity turned in the direction of its propagation.

the sound of a cataract; it is able sensibly to extinguish the tinkle of the cow-bells. Still, as I have said, the sound-shadow is but partial, and the marker at the rifle-butts never fails to hear the explosion, though he is well protected from the ball. The most striking example of this inflection of a sonorous wave that I have ever seen, was exhibited at Erith after the tremendous explosion of a powder magazine which occurred there in 1864. The village of Erith was some miles distant from the magazine, but in nearly all cases the windows were shattered; and it was noticeable that the windows turned away from the origin of the explosion suffered almost as much as those which faced it. Lead sashes were employed in Erith church, and these being in some degree flexible, enabled the windows to yield to pressure without much fracture of the glass. Every window in the church, front and back, was bent *inwards*. In fact, as the sound-wave reached the church it separated right and left, and, for a moment, the edifice was clasped by a girdle of intensely compressed air, which forced all its windows inwards. After compression, the air within the church no doubt dilated, and tended to restore the windows to their first condition. The bending in of the windows, however, produced but a small condensation of the whole mass of air within the church; the force of recoil was therefore feeble in comparison with the force of impact, and insufficient to undo what the latter had accomplished.

Having thus dealt with the reflection, the refraction, and the inflection of sound, we must fix our attention more closely than we have hitherto done on the two conditions which determine the velocity of propagation of a sonorous wave; namely, the elasticity and the density of the medium through which the wave passes. The elasticity of air is measured by the pressure which it sustains or can hold in equilibrium. At the sea-level this pressure is equal to that of a stratum of mercury about 30 inches high. At the summit of Mont Blanc the barometric column is not much more than half this height; and, consequently, the elasticity of the air upon the summit of the mountain is not much more than half what it is at the sea-level.

If we could augment the elasticity of air, without at the same time augmenting its density, we should augment the velocity of sound. Or, if allowing the elasticity to remain constant we could diminish the density, we should augment the velocity of sound. Now, air in a closed vessel, where it cannot expand, has its elasticity augmented by heat, while its density remains unchanged. Through such heated air sound travels more rapidly than through cold air. Again, air free to expand has its density lessened by warming, its elasticity remaining the same, and through such air sound travels more rapidly than through cold air. This is the case with our atmosphere when heated by the sun. It expands and becomes lighter, bulk for bulk, while its pressure, or in other words its elasticity, remains the same. And now you see the reason of the phrase that the velocity of sound in air '*at the freezing temperature*' is 1,090 feet a second. At all lower temperatures the velocity is less than this, and at all higher temperatures it is greater. The late M. Wertheim has determined the velocity of sound in air of different temperatures, and here are some of his results:—

Temperature of air.	Velocity of sound.
0·5° centigrade . .	1089 feet
2·10 " . .	1091 "
8·5 " . .	1109 "
12·0 " . .	1113 "
26·6 " . .	1140 "

At a temperature of half a degree above the freezing point of water the velocity is 1,089 feet a second; at a temperature of 26·6 degrees, it is 1,140 feet a second, or a difference of 51 feet for 26 degrees, that is to say, an augmentation of velocity of about 2 feet for every single degree centigrade.

With the same elasticity the density of hydrogen gas is much less than that of air, and the consequence is that the velocity of sound in hydrogen far exceeds its velocity in air. The reverse holds good for heavy carbonic acid gas. Here, while the elasticity is only equal to that of air, the density is greater, and consequently the velocity of sound less. If density and elasticity vary in the same proportion, as the law of Mariotte proves them to do in air, when the temperature is preserved constant, they neutralise each other's effect; hence, if the temperature were the same, the velocity of sound upon the summits of the highest Alps would be the same as at the mouth of the Thames. But inasmuch as the air above is colder than that below, the actual velocity on the summits of the mountains is less than that at the sea-level. To express this result in stricter language, the velocity is *directly* proportional to the square root of the elasticity of the air; it is also *inversely* proportional to the square root of the density of the air. Consequently, as in air of a constant temperature elasticity and density vary in the same proportion, and act oppositely, the velocity of sound is not affected by a change of density, if unaccompanied by a change of temperature.

There is no mistake more common than to suppose the velocity of sound to be augmented by density. The mistake has arisen from a misconception of the fact, that in solids and liquids the velocity is greater than in gases. But it is the high elasticity of those bodies, *in relation to their density*, that causes sound to pass rapidly through them. Other things remaining the same, an augmentation of density always produces a diminution of velocity. Were the elasticity of water, which is measured by its compressibility, only equal to that of air, the velocity of sound in water, instead of being more than quadruple the velocity in air, would be only a small fraction of that velocity. Both density and elasticity, then, must be always borne in mind; the velocity of sound being determined by neither taken separately, but by the relation of the one to the other. The effect of small density and high elasticity is exemplified in an astonishing manner by the luminiferous ether, which transmits the vibrations of light—not at the rate of so many feet, but at the rate of nearly two hundred thousand miles in a second of time.

It is now time to say a word about the manner in which the velocity of sound in air has been ascertained. Those who are unacquainted with the details of scientific investigation have no idea of the amount of labour expended in the determination of those numbers on which important calculations or inferences depend. They have no idea of the patience shown by a Berzelius in determining atomic weights; by a Regnault in determining coefficients of expansion; or by a Joule in determining the mechanical equivalent of heat. There is a morality brought to bear upon such matters which, in point of severity, is probably without a parallel in any other domain of intellectual action. The desire for anything but the truth must be absolutely annihilated; and to attain perfect accuracy no labour must be shirked, no difficulty ignored. Thus, as regards the determination of the velocity of sound in air, hours might be filled with a simple statement of the efforts made to establish it with precision. The question has occupied the attention of philosophers in England, France, Germany, Italy, and Holland. But to the French and Dutch philosophers we owe the application of the last refinements of experimental skill to the solution of the problem. They neutralised effectually the influence of the wind; they took into account barometric pressure, temperature, and hygrometric condition. Sounds were started at the same moment from two distant stations, and thus caused to travel from station to station through the selfsame air. The distance between the stations was determined by exact trigonometrical observations, and means were devised for measuring with the utmost accuracy the time required by the sound to pass from the one station to the other. This time, expressed in seconds, divided into the distance expressed in feet, gave 1,090 feet per second as the velocity of sound through air at the temperature of 0° centigrade.

The time required by light to travel over all terrestrial distances is practically zero; and in the experiments just referred to the moment of explosion was marked by the flash of a gun, the time occupied by the sound in passing from station to station being the interval observed between the appearance of the flash and the arrival of the sound. The velocity of sound in air once established, it is plain that we can apply it to the determination of distances. By observing, for example, the interval between the appearance of a flash of lightning and the arrival of the accompanying thunder peal, we at once determine the distance of the place of discharge. It is only when the interval between the flash and peal is short that danger from lightning is to be apprehended.

I now come to one of the most delicate points in the whole theory of sound. The velocity through air has been determined by direct experiment; but knowing the elasticity and density of the air, it is possible without any experiment at all to calculate the velocity with which a sound-wave is transmitted through it. Sir Isaac Newton made this calculation, and found the velocity at the freezing temperature to be 916 feet a second. This is about one-sixth less than actual observation had proved the velocity to be, and the most curious suppositions were made to account for the discrepancy. Newton himself threw out the conjecture that it was only in passing from particle to particle of the air that sound required *time* for its transmission; that it moved instantaneously *through the particles themselves.* He then supposed the line along which sound passes to be occupied by air-particles for one-sixth of its extent, and thus he sought to make good the missing velocity. The very art and ingenuity of this assumption were sufficient to ensure its rejection; other theories were therefore advanced, but the great French mathematician Laplace was the first to completely solve the enigma. I shall now endeavour to make you thoroughly acquainted with his solution.

Fig. 10.

I hold in my hand a strong cylinder of glass, T U, fig. 10, accurately bored, and quite smooth within. Into this cylinder, which is closed at the bottom, fits this air-tight piston. By pushing the piston down, I condense the air beneath it; and when I do so heat is developed. Attaching a scrap of amadou to the bottom of the piston, I can ignite it by the heat generated by compression. Dipping a bit of cotton wool into bisulphide of carbon, and attaching it to the piston, when the latter is forced down, a flash of light, due to the ignition of the bisulphide of carbon vapour, is observed within the tube. It is thus proved that when air is compressed, heat is generated. By another experiment, I can show you that when air is rarefied, cold is developed. This brass box contains a quantity of condensed air. I open the cock, and permit the air to discharge itself against a suitable thermometer; the sinking of the instrument declares the chilling of the air.

All that you have heard regarding the transmission of a sonorous pulse through air, is, I trust, still fresh in your minds. As the pulse advances it squeezes the particles of air together, and two results follow from this compression of the air. Firstly, its elasticity is augmented through the mere augmentation of its density. Secondly, its elasticity is augmented by the heat developed by compression. It was the change of elasticity which resulted from a change of density that Newton took into account, and he entirely overlooked the augmentation of elasticity due to the second cause above mentioned. Over and above, then, the elasticity involved in Newton's calculation, we have an additional elasticity due to changes of temperature produced by the sound itself. When both are taken into account, the calculated and the observed velocity agree perfectly.

But here, without due caution, we may fall into the gravest error. In fact, in dealing with nature, the mind must be on the alert to seize all her conditions; otherwise we soon learn that our thoughts are not in accordance with her facts. Now the augmentation of velocity due to the changes of temperature produced by the sonorous wave itself, is totally different from the augmentation which arises from the heating of the general mass of the air. The *average* temperature of the air is unchanged by the waves of sound. We cannot have a condensed pulse without having a rarefied one associated with it. But in the rarefaction the temperature of the air is as much lowered as it is raised in the condensation. Supposing then the atmosphere parcelled out into such condensations and rarefactions, with their respective temperatures, an extraneous sound passing through such an atmosphere would be as much retarded in the latter as accelerated in the former, and no variation of the average velocity could arise from such a distribution of temperature.

Whence then does the augmentation pointed out by Laplace arise? I would ask your best attention while I endeavour to make this knotty point clear to you. If air be compressed it becomes smaller in volume; if the pressure be diminished the volume expands. The force which resists compression, and which produces expansion, is the elastic force of the air. Thus an external pressure squeezes the air-particles together; their own elastic force holds them asunder, and the particles are in equilibrium when these two forces are in equilibrium. Hence it is that the external pressure is a measure of the elastic force. Let the middle row of dots, fig. 11, represent a series of air-particles in a state of quiescence between the points a and x. Then, because of the elastic force exerted between the particles, if any one of them be moved from its position of rest, the motion will be transmitted through the

FIG. 11.

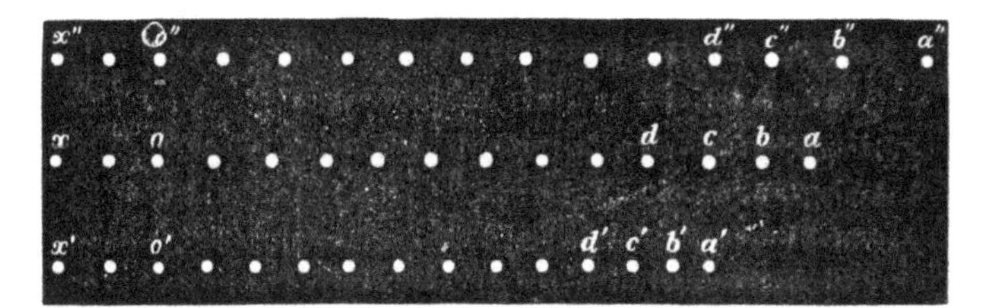

entire series. Supposing, then, the particle a to be driven by the prong of a tuning-fork, or some other vibrating body, towards x, so as to be caused finally to occupy the position a' in the lowest row of particles. At the instant the excursion of a commences, its motion begins to be transmitted to b. In the next following moments b transmits the motion to c; c to d; d to e, and so on. So that by the time a has reached a', the motion will have been propagated to some point o' of the line of particles more or less distant from a'. The entire series of particles between a' and o' is then in a state of condensation. The distance a' o', over which the motion has travelled during the excursion of a to a', will depend upon the elastic force exerted between the particles. Fix your attention on any two of the particles, say *a* and *b*. The elastic force between them may be figured as a spiral spring, and it is plain that the more flaccid this spring the more sluggish would be the communication of motion from *a* to *b* ; while the stiffer the spring the more prompt would be the communication of motion. What is true of *a* and *b* is true for every other pair of particles between *a* and *o*. Now the spring between every pair of these particles is stiffened by the heat developed along the line of condensation, and hence the velocity of propagation is augmented by this heat. Reverting to our old experiment with the row of boys, it is as if, by the very act of pushing his neighbour, the muscular rigidity of each boy's arm was increased, thus enabling him to deliver his push more promptly than he would have done without this increase of rigidity. The condensed portion of a sonorous wave is propagated in the manner here described, and it is plain that the velocity of propagation is augmented by the heat developed in the condensation.

Let us now turn our thoughts for a moment to the propagation of the rarefaction. Supposing, as before, the middle row *a x* to represent the particles of air in equilibrium under the pressure of the atmosphere ; and suppose the particle *a* to be suddenly drawn to the right, so as to occupy the position a'' in the highest line of dots: a'' is immediately followed by b'', b'' by c'', c'' by d'', d'' by e''; and thus the rarefaction is propagated backward towards x'', reaching a point o'' in the line of particles by the time *a* has completed its motion to the right. Now, why does b'' follow a'', when a'' is drawn away from it? Manifestly because the elastic force exerted between b'' and a'' is less than that between b'' and c''. In fact, b'' will be driven after a'' by a force equal to the difference of the two elasticities between a'' and b'' and between b'' and c''. The same remark applies to the motion of c'' after b'', to that of d'' after c'', in fact, to the motion of each succeeding particle when it follows its predecessor. The greater the difference of elasticity on the two sides of any particle the more promptly will it follow its predecessor. And here see what the cold of rarefaction accomplishes. In addition to the diminution of the elastic force between a'' and b'' by the withdrawal of a'' to a greater distance, there is also a further diminution due to the lowering of the temperature. The cold developed augments the difference of elastic force on which the propagation of the rarefaction depends. Thus we see, that because the heat developed in the condensation augments the rapidity of the condensation, and because the cold developed in the rarefaction augments the rapidity of the rarefaction, the sonorous wave, which consists of a condensation and a rarefaction, must have its velocity augmented by the heat *and* the cold which it develops during its own progress.

It is worth while fixing your attention here upon the fact that the distance $a'\,o'$ to which the motion has been propagated while a is moving to a' may be vastly greater than that passed over in the same time by the particle itself. The excursion of a' may not be more than a small fraction of an inch, while the distance to which the motion is transferred during the time required by a' to perform this small excursion may be many feet, or even many yards. If this point should not appear altogether plain to you now, it will appear so by-and-by.

Having grasped this, even partially, I will ask you to accompany me to a remote corner of the domain of natural philosophy, with the view, however, of showing that remoteness does not imply discontinuity. Let a certain quantity of air at a temperature of 0°, contained in a perfectly inexpansible vessel, have its temperature raised 1°. Let the same quantity of air, placed in a vessel which permits the air to expand when it is heated,—the pressure on the air being kept constant during its expansion,—also have its temperature raised 1°. The quantities of heat employed in the two cases are different. The one quantity expresses what is called the specific heat of air at constant volume, the other the specific heat of air at constant pressure.* It is an instance of the manner in which apparently unrelated natural phenomena are bound together, that from the calculated and observed velocities of sound in air, we can deduce the ratio of these two specific heats. In fact, if the calculated and observed velocities be both squared, then dividing the greater square by the less we obtain the ratio referred to. Calling the specific heat at constant volume C^v, and that at constant pressure C^p; calling, moreover, Newton's calculated velocity V, and the observed velocity V′, Laplace proved that

$$\frac{C^p}{C^v} = \frac{V'^2}{V^2}$$

Inserting the values of V and V′ in this equation, and making the calculation, we find

$$\frac{C^p}{C^v} = 1{\cdot}42.$$

Thus, without knowing either the specific heat at constant volume or at constant pressure, Laplace found the ratio of the greater of them to the less to be 1·42. It is evident from the foregoing formula that the calculated velocity of sound, multiplied by the square root of this ratio, gives the observed velocity.

But there is one assumption connected with the determination of this ratio, which must be here brought clearly forth. It is assumed that the heat developed by compression *remains in the condensed portion of the wave*, and applies itself there to augment the elasticity; that no portion of the heat is lost by radiation. If air were

* See *Heat as a Mode of Motion*, 2nd edition, pp. 69 to 72.

a powerful radiator, this assumption could not stand. The heat developed in the condensation could not, then, remain in the condensation. It would radiate all round, lodging itself for the most part in the chilled and rarefied portion of the wave, which would be gifted with a proportionate power of absorption. Hence the direct tendency of radiation would be to equalise the temperatures of the different parts of the wave, and thus to abolish the augmentation of velocity which called forth Laplace's correction.*

The question, then, of the correctness of this ratio involves the other, and apparently incongruous question, whether atmospheric air possesses any sensible radiative power. If the ratio be correct, the practical absence of radiative power on the part of air is demonstrated. How then are we to ascertain whether the ratio is correct or not? By a process of reasoning which illustrates still further how natural agencies are intertwined. It was this ratio, looked at by a man of genius, and I may add a man of suffering, named Mayer, which helped him to a clearer and a grander conception of the relation and interaction of the forces of inorganic and organic nature, than any philosopher up to his time had attained. Mayer was the first to see in this ratio an illustration of the destruction of heat. He was the first to see that the excess 0·42 of the specific heat at constant pressure over that at constant volume was the quantity of heat consumed in the work performed by the expanding gas. And, supposing the air to be confined laterally and permitted to expand in a vertical direction, in which direction it would simply have to lift the weight of the atmosphere, Mayer attempted to calculate the precise amount of heat consumed in the raising of

* In fact, the prompt abstraction of the motion of heat from the condensation, and its prompt communication to the rarefaction by the contiguous luminiferous ether, would prevent the former from ever rising so high, or the latter from ever falling so low, in temperature as it would do if the power of radiation and absorption was absent.

this or any other weight. He thus sought to determine the 'mechanical equivalent' of heat. In the combination of his data his mind was as clear as the day, but for the numerical correctness of these data he was obliged to rely upon the experimenters of his age. Their results, though approximately correct, were not *so* correct as the transcendent experimental ability of Regnault, aided by the last refinements of constructive skill, afterwards made them. Without changing in the slightest degree the method of his thought, or the structure of his calculation, the simple introduction of the exact numerical data into the formula of Mayer brings out the true mechanical equivalent of heat.

But how are we able to speak thus confidently of the accuracy of this equivalent? We are enabled to do so by the labours of an Englishman, who worked at this subject contemporaneously with Mayer; and who, while animated by the creative genius of his celebrated German brother, enjoyed also the opportunity of bringing the inspirations of that genius to the test of experiment. By the immortal experiments of Mr. Joule, the mutual convertibility of mechanical work and heat was first conclusively established. And 'Joule's equivalent,' as it is rightly called, considering the amount of labour and skill expended in its determination, is almost identical with that derived from the formula of Mayer.

Consider now the ground over which we have trodden, the curious labyrinth of reasoning and experiment through which we have passed. We started with the observed and calculated velocities of sound in atmospheric air. We found Laplace, by a special assumption, deducing from these velocities the ratio of the specific heat of air at constant pressure, to its specific heat at constant volume. We found Mayer calculating from this ratio the mechanical equivalent of heat; finally, we found Mr. Joule determining the same equivalent by direct experiments on the friction of solids and liquids. And what is the result? Mr. Joule's experiments prove the result of Mayer to be the true one; they therefore prove the ratio determined by Laplace to be the true ratio; and, because they do this, they prove at the same time the practical absence of radiative power in atmospheric air. It seems a long step from the stirring of water, or the rubbing together of iron plates, to the radiation of the atoms of our atmosphere; both questions are, however, connected by the line of reasoning here followed out.

But the true physical philosopher never rests content with an inference when an experiment to verify or contravene it is possible. The foregoing argument is clenched by bringing the radiative power of atmospheric air to an experimental test. When this is done, experiment and reasoning are found to agree; air being proved to be a body sensibly devoid of radiative and absorptive power.

But here I think the experimenter on the transmission of sound through gases needs a word of warning. In Laplace's day, and long subsequently, it was thought that gases of all kinds possessed only an infinitesimal power of radiation; but that this is not the case is now well established. It would, I think, be rash to assume that, in the case of such bodies as ammonia, aqueous vapour, sulphurous acid, and olefiant gas, their enormous radiative powers do not interfere with the application of the formula of Laplace. It behoves us to enquire whether the ratio of the two specific heats deduced from the velocity of sound in these bodies is the true ratio; and whether, if the true ratio could be found by other methods, its square root, multiplied into the calculated velocity, would give the observed velocity. From the moment heat first appears in the condensation and cold in the rarefaction of a sonorous wave in any of those gases, the radiative power comes into play to abolish the difference of temperature. The condensed part of the wave is on this account rendered more flaccid, and the rarefied part less flaccid than it would otherwise be, and with a sufficiently high radiative power the velocity of sound, instead of coinciding with that derived from the formula of Laplace, must approximate to that derived from the more simple formula of Newton.

To complete our knowledge of the transmission of sound through gases, I add a table from the excellent researches of Dulong, who employed in his experiments a method which shall be subsequently explained.

VELOCITY OF SOUND IN GASES AT THE TEMPERATURE OF 0° C.

	Velocity.
Air	1,092 feet.
Oxygen	1,040 „
Hydrogen	4,164 „
Carbonic acid	858 „
Carbonic oxide	1,107 „
Protoxide of nitrogen	859 „
Olefiant gas	1,030 „

According to theory, the velocities of sound in oxygen and hydrogen are inversely proportional to the square roots of the densities of the two gases. We here find this theoretic deduction verified by experiment. Oxygen being sixteen times heavier than hydrogen, the velocity of sound in the latter gas ought, according to the above law, to be four times its velocity in the former; hence the velocity in oxygen being 1,040, in hydrogen calculation would make it 4,160. Experiment, we see, makes it 4,164.

The velocity of sound in liquids may be determined theoretically, as Newton determined its velocity in air; for the density of a liquid is easily determined, and its elasticity can be measured by subjecting it to compression. In the case of water the calculated and the observed velocities agree so closely as to prove that the changes of temperature produced by a sound-wave in water, have no sensible influence upon the velocity. In a series of memorable experiments in the lake of Geneva, MM. Colladon and Sturm determined the velocity of sound through water, and made it 4,708 feet a second. By a mode of experiment which you will subsequently be able to comprehend, the late M. Wertheim determined the velocity through various liquids, and in the following table I have collected his results:—

TRANSMISSION OF SOUND THROUGH LIQUIDS.

Name of Liquid	Temperature	Velocity
		feet
River water (Seine)	15° C.	4,714
" "	30	5,013
" "	60	5,657
Sea water (artificial)	20	4,768
Solution of common salt	18	5,132
Solution of sulphate of soda . . .	20	5,194
Solution of carbonate of soda . . .	22	5,230
Solution of nitrate of soda	21	5,477
Solution of chloride of calcium . . .	23	6,493
Common alcohol	20	4,218
Absolute alcohol	23	3,804
Spirits of turpentine	24	3,976
Sulphuric ether	0	3,801

We learn from this table that sound travels with different velocities through different liquids; that a salt dissolved in water augments the velocity, and that the salt which produces the greatest augmentation is chloride of calcium. The experiments also teach us that in water, as in air, the velocity augments with the temperature. At a temperature of 15° C., for example, the velocity in Seine water is 4,714 feet, at 30° it is 5,013 feet, and at 60° 5,657 feet a second.

I have said that from the compressibility of a liquid, determined by proper measurements, the velocity of sound through the liquid may be deduced. Conversely from the velocity of sound in a liquid, the compressibility of the liquid may be deduced. Wertheim compared a series of compressibilities deduced from his experiments on sound, with a similar series obtained directly by M. Grassi. The agreement of both, exhibited in the following table, is a strong confirmation of the accuracy of the method pursued by Wertheim:—

	Cubic compressibility	
	from Wertheim's velocity of sound.	from the direct experiments of M. Grassi.
Sea water	0·0000467	0·0000436
Solution of common salt .	0·0000349	0·0000321
" carbonate of soda	0·0000337	0·0000297
" nitrate of soda .	0·0000301	0·0000295
Absolute alcohol . . .	0·0000947	0·0000991
Sulphuric ether . . .	0·0001002	0·0001110

The greater the resistance which a liquid offers to compression, the more promptly and forcibly will it return, after it has been compressed, to its original volume. The less the compressibility, therefore, the greater is the elasticity, and consequently, other things being equal, the greater the velocity of sound through the liquid.

We have now to examine the transmission of sound through solids. Here, as a general rule, the elasticity, as compared with the density, is greater than in liquids, and consequently the propagation of sound is more rapid. In the following table the velocity of sound through various metals, as determined by Wertheim, is recorded:—

VELOCITY OF SOUND THROUGH METALS.

Name of Metal	at 20° C.	at 100° C.	at 200° C.
Lead	4,030	3,951	—
Gold	5,717	5,640	5,691
Silver	8,553	8,658	8,127
Copper	11,666	10,802	9,690
Platinum	8,815	8,437	8,079
Iron	16,822	17,386	15,483
Iron wire (ordinary) . . .	16,130	16,728	—
Cast steel	16,357	16,153	15,709
Steel wire (English) . . .	15,470	17,201	16,394
Steel wire	16,023	16,443	—

The foregoing table illustrates the influence of temperature on the velocity of sound through metals. As a general rule, the velocity is diminished by augmented temperature; iron is, however, a striking exception to this rule, but it is only within certain limits an exception. While, for example, a rise of temperature from 20° to 100° C. in the case of copper causes the velocity to fall from 11,666 to 10,802, the same rise produces in the case of iron an increase of velocity from 16,882 to 17,386. Between 100° and 200°, however, we see that iron falls from the last figure to 15,483. In iron, therefore, up to a certain point, the elasticity is augmented by heat; beyond that point it is lowered. Silver is also an example of the same kind.

The difference of velocity in iron and in air may be illustrated by the following instructive experiment. Choose one of the longest horizontal bars employed for fencing in Hyde Park; and let an assistant strike the bar at one end while the ear of the observer is held close to the bar at a considerable distance. Two sounds will reach the ear in succession; the first being transmitted through the iron and the second through the air. This effect was obtained by M. Biot, in his experiments on the iron water-pipes of Paris.

The transmission of sound through any body depends in some degree on the manner in which the molecules of the body are arranged. If the body be homogeneous and without structure, sound is transmitted through it equally well in all directions. But this is not the case when the body, whether unorganised like a crystal, or organised like a tree, possesses a definite structure. This is also true of other things than sound. Subjecting, for example, a sphere of wood to magnetic force, it is not equally affected in all directions. It is repelled by the pole of a magnet, but it is most strongly repelled when the force acts along the fibre. Heat also is conducted with different facilities in different directions through wood. It is most freely conducted along the fibre; but across the fibre the power of wood to conduct heat is not the same in all directions. It passes more freely across the ligneous layers than along them. Wood, therefore, possesses three unequal axes of calorific conduction. These coincide with the axes of elasticity discovered by Savart. MM. Wertheim and Chevandier have determined the velocity of sound along these three axes and obtained the following results:—

VELOCITY OF SOUND IN WOOD.

Name of Wood	Along Fibre	Across Rings	Along Rings
Acacia	15,467	4,840	4,436
Fir	15,218	4,382	2,572
Beech	10,965	6,028	4,643
Oak	12,622	5,036	4,229
Pine	10,900	4,611	2,605
Elm	13,516	4,665	3,324
Sycamore	14,639	4,916	3,728
Ash	15,314	4,567	4,142
Alder	15,306	4,491	3,423
Aspen	16,677	5,297	2,987
Maple	13,472	5,047	3,401
Poplar	14,050	4,600	3,444

Thus, separating a cube from the bark-wood of a good-sized tree, where the rings for a short distance may be regarded as straight; if A R, fig. 12, be the section of the tree, then the velocity of the sound in direction *m n*, through such a cube, is greater than in the direction *a b*.

FIG. 12.

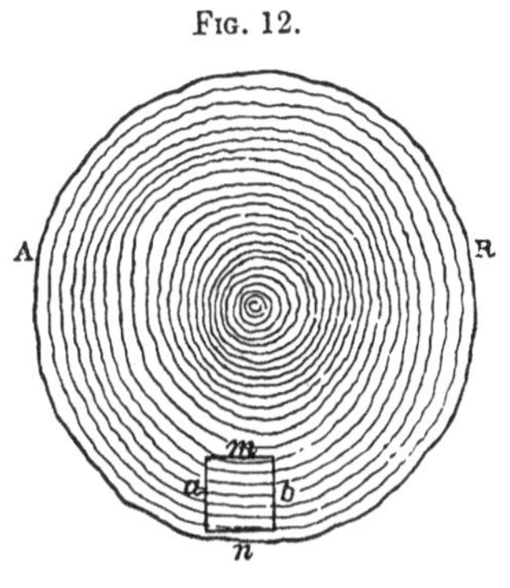

The foregoing table strikingly illustrates the influence of molecular structure. The great majority of crystals show differences of the same kind. Such bodies, for the most part, have their molecules arranged in different degrees of proximity in different directions, and where this occurs there are sure to be differences in the transmission and manifestation of heat, light, electricity, magnetism, and sound.

I will conclude this lecture on the transmission of sound through gases, liquids, and solids, by a quaint and beautiful extract from the writings of that admirable thinker, Dr. Robert Hooke; and I do so the more willingly, because I fear his contiguity to Newton has too much dimmed the name and fame of this extraordinary man. It will be noticed that the philosophy of the stethoscope is enunciated in the following passage, and I am hardly acquainted with another which illustrates so well that action of the scientific imagination which, in all great investigators, is the precursor and associate of experiment.

'There may also be a possibility,' writes Hooke, 'of discovering the internal motions and actions of bodies by the sound they make. Who knows but that, as in a watch, we may hear the beating of the balance, and the running of the wheels, and the striking of the hammers, and the grating of the teeth, and multitudes of other noises; who knows, I say, but that it may be possible to discover the motions of the internal parts of bodies, whether animal, vegetable, or mineral, by the sound they make, that one may discover the works performed in the several offices and shops of a man's body, and thereby discover what instrument or engine is out of order, what works are going on at several times, and lie still at others, and the like; that in plants and vegetables one might discover by the noise the pumps for raising the juice, the valves for stopping it, and the rushing of it out of one passage into another, and the like? I could proceed further, but methinks I can hardly forbear to blush, when I consider how the most part of men will look upon this; but, yet again, I have this encouragement, not to think all these things utterly impossible, though never so much derided by the generalit of men, and never so seemingly mad, foolish, and phantastic, that, as the thinking them impossible cannot much improve my knowledge, so the believing them possible may perhaps be an occasion of taking notice of such things as another would pass by without regard as useless. And somewhat more of encouragement I have also from experience, that I have been able to hear very plainly the beating of a man's heart, and 'tis common to hear the motion of wind to and fro in the guts, and other small vessels; the stopping of the lungs is easily discovered by the wheesing, the stopping of the head by the humming and whistling noises, the slipping to and fro of the joints, in many cases, by crackling, and the like, as to the working or motion of the parts one amongst another; methinks I could receive encouragement from hearing the hissing noise made by a corrosive menstruum in its operation, the noise of fire in dissolving, of water in boyling, of the parts of a bell after that its motion is grown quite invisible as to the eye, for to me these motions and the other seem only to differ *secundum magis minus*, and so to their becoming sensible they require either that their motions be increased, or that the organ be made more nice and powerful to sensate and distinguish them.'

SUMMARY OF LECTURE I.

The sound of an explosion is propagated as a wave or pulse through the air.

This wave impinging upon the tympanic membrane causes it to shiver, its tremors are transmitted to the auditory nerve, and along the auditory nerve to the brain, where it announces itself as sound.

A sonorous wave consists of two parts, in one of which the air is condensed, and in the other rarefied.

The motion of the sonorous wave must not be confounded with the motion of the particles which at any moment form the wave. During the passage of the wave every particle concerned in its transmission makes only a small excursion to and fro.

The length of this excursion is called the *amplitude* of the vibration.

Sound cannot pass through a vacuum.

Sound is in all respects reflected like light; it is also refracted like light; and it may, like light, be condensed by suitable lenses.

Sound is also inflected, the sonorous wave bending round obstacles; such obstacles, however, in part shade off the sound.

Echoes are produced by the reflected waves of sound.

In regard to sound and the medium through which it passes, four distinct things are to be borne in mind:—intensity, velocity, elasticity, and density.

The intensity is proportional to the square of the amplitude as above defined.

It is also proportional to the square of the maximum velocity of the vibrating air-particles.

When sound issues from a small body in free air, the intensity diminishes as the square of the distance from the body increases.

If the wave of sound be confined in a tube with a smooth interior surface, it may be conveyed to great distances without sensible loss of intensity.

The velocity of sound in air depends on the elasticity of the air in relation to its density. The greater the elasticity the swifter is the propagation; the greater the density the slower is the propagation.

The velocity is directly proportional to the square root of the elasticity; it is inversely proportional to the square root of the density.

Hence, if elasticity and density vary in the same proportion, the one will neutralise the other as regards the velocity of sound.

That they do vary in the same proportion is proved by the law of Mariotte; hence, the velocity of sound in air is independent of the density of the air.

But that this law shall hold good, it is necessary that the dense air and the rare air should have the same temperature.

The intensity of a sound depends upon the density of the air in which it is generated, but not on that of the air in which it is heard.

The velocity of sound in air of the temperature 0° C. is 1,090 feet a second; it augments about 2 feet for every degree centigrade added to its temperature.

Hence, given the velocity of sound in air, the temperature of the air may be readily calculated.

The distance of a fired cannon or of a discharge of lightning may be determined by observing the interval which elapses between the flash and the sound.

From the foregoing, it is easy to see that if a row of soldiers form a circle, and discharge their pieces all at the same time, the sound will be heard as a single discharge by a person occupying the centre of the circle.

But if the men form a straight row, and if the observer stand at one end of the row, the simultaneous discharge of the men's pieces will be prolonged to a kind of roar.

A discharge of lightning along a lengthy cloud may in this way produce the prolonged roll of thunder. The roll of thunder, however, must in part at least be due to echoes from the clouds.

The pupil will find no difficulty in referring many common occurrences to the fact that sound requires a sensible time to pass through any considerable length of air. For example, the fall of the axe of a distant woodcutter is not simultaneous with the sound of the stroke. A company of soldiers marching to music along a road cannot march in time, for the notes do not reach those in front and those behind simultaneously.

In the condensed portion of a sonorous wave the air is above, in the rarefied portion of the wave it is below, its average temperature.

This change of temperature, produced by the passage of the sound-wave itself, virtually augments the elasticity of the air, and makes the velocity of sound about $\frac{1}{6}$th greater than it would be if there were no change of temperature.

The velocity found by Newton, who did not take this change of temperature into account, was 916 feet a second.

Laplace proved that by multiplying Newton's velocity by the square root of the ratio of the specific heat of air at constant pressure, to its specific heat at constant volume, the actual or observed velocity is obtained.

Conversely, from a comparison of the calculated and observed velocities, the ratio of the two specific heats may be inferred.

The mechanical equivalent of heat may be deduced from this ratio; it is found to be the same as that established by direct experiment.

This coincidence leads to the conclusion that atmospheric air is devoid of any sensible power to radiate heat. Direct experiments on the radiative power of air establish the same result.

The velocity of sound in water is more than four times its velocity in air.

The velocity of sound in iron is seventeen times its velocity in air.

The velocity of sound along the fibre of pine-wood is ten times its velocity in air.

The cause of this great superiority is that the elasticities of the liquid, the metal, and the wood, as compared with their respective densities, are vastly greater than the elasticity of air in relation to its density.

The velocity of sound is dependent to some extent upon molecular structure. In wood, for example, it is conveyed with different degrees of rapidity in different directions.

LECTURE II.

PHYSICAL DISTINCTION BETWEEN NOISE AND MUSIC—A MUSICAL TONE PRODUCED BY PERIODIC, NOISE PRODUCED BY UNPERIODIC, IMPULSES—PRODUCTION OF MUSICAL SOUNDS BY TAPS—PRODUCTION OF MUSICAL SOUNDS BY PUFFS—DEFINITION OF PITCH IN MUSIC—VIBRATIONS OF A TUNING-FORK; THEIR GRAPHIC REPRESENTATION ON SMOKED GLASS—OPTICAL EXPRESSION OF THE VIBRATIONS OF A TUNING-FORK—DESCRIPTION OF THE SYREN—LIMITS OF THE EAR; HIGHEST AND DEEPEST TONES—RAPIDITY OF VIBRATION DETERMINED BY THE SYREN—DETERMINATION OF THE LENGTHS OF SONOROUS WAVES—WAVE-LENGTHS OF THE HUMAN VOICE IN MAN AND WOMAN—TRANSMISSION OF MUSICAL SOUNDS THROUGH LIQUIDS AND SOLIDS.

IN our last lecture we considered the propagation of a single sonorous wave through air, the origin of the wave being the explosion of a small balloon filled with a mixture of oxygen and hydrogen. A sound of momentary duration was thus generated. We have to-day to consider continuous sounds, and to make ourselves in the first place acquainted with the physical distinction between noise and music. As far as sensation goes, every body knows the difference between these two things. But we have now to enquire into the causes of sensation, and to make ourselves acquainted with that condition of the external air which in one case resolves itself into music and in another into noise.

We have already learned that what is loudness in our sensations is outside of us nothing more than width of swing, or *amplitude*, of the vibrating air-particles. Every other real sonorous impression of which we are conscious, has its correlative without, as a mere form or state of the atmosphere. Could we, for example, see the air through which the sound of an agreeable voice is passing, we might see stamped upon that air the conditions of motion on which the sweetness of the voice depends. In ordinary conversation, also, the physical precedes and arouses the psychological; the spoken language, which is to give us pleasure or pain, which is to rouse us to anger or soothe us to peace, existing for a time, between us and the speaker, as a purely mechanical condition of the intervening air.

If I shake this tool-box, with its nails, bradawls, chisels, and files, you hear what we should call noise. If I draw a violin bow across this tuning-fork, you hear what we should call music. The noise affects us as an irregular succession of shocks. We are conscious while listening to it of a jolting and jarring of the auditory nerve, while the musical sound flows smoothly and without asperity or irregularity. How is this smoothness secured? *By rendering the impulses received by the tympanic membrane perfectly periodic.* A periodic motion is one that repeats itself. The motion of a common pendulum, for example, is periodic, and as it swings through the air it produces waves or pulses which follow each other with perfect regularity. Such waves, however, are far too sluggish to excite the auditory nerve. To produce a musical tone we must have a body which vibrates with the unerring regularity of the pendulum, but which can impart much sharper and quicker shocks to the air.

Imagine the first of a series of pulses which follow each other at regular intervals, impinging upon the tympanic membrane. It is shaken by the shock; and a body once shaken cannot come instantaneously to rest. The human ear, indeed, is so constructed that the sonorous motion vanishes with extreme rapidity, but its disappearance is not instantaneous; and if the motion imparted to the auditory nerve by each individual pulse of our series continue until the arrival of its successor, the sound will not cease at all. The effect of every shock will be renewed before it vanishes, and the recurrent impulses will link themselves together to a continuous musical sound. The pulses on the contrary which produce noise are of irregular strength and recurrence. They dash confusedly into the ear, and reproduce their own unpleasant confusion in our sensations. Music resembles poetry of smooth and perfect rhythm, noise resembles harsh and rumbling prose. But as the words of the prose might, by proper arrangement, be reduced to poetry, so also by rendering its elements periodic the uproar of the streets might be converted into the music of the orchestra. The action of noise upon the ear has been well compared to that of a flickering light upon the eye, both being painful through the sudden and abrupt changes which they impose upon their respective nerves.

The only condition necessary to the production of a musical sound is that the pulses should succeed each other in the same interval of time. No matter what its origin may be, if this condition be fulfilled the sound becomes musical. If a watch, for example, could be caused to tick with sufficient rapidity—say one hundred times a second—the ticks would lose their individuality and blend to a musical tone. And if the strokes of a pigeon's wings could be accomplished at the same rate, the progress of the bird through the air would be accompanied by music. In the humming bird the necessary rapidity is attained; and when we pass on from birds to insects, where the vibrations are more rapid, we have a musical note as the ordinary accompaniment of the insects' flight.* The puffs of a locomotive at starting follow each other slowly at first, but they soon increase so

* According to Burmeister, through the injection and ejection of air into and from the cavity of the chest.

rapidly as to be almost incapable of being counted. If this increase could continue until the puffs numbered 50 or 60 a second, the approach of the engine would be heralded by an organ peal of tremendous power.

Galileo produced a musical sound by passing a knife over the edge of a piastre. The minute serration of the coin indicated the periodic character of the motion, which consisted of a succession of taps quick enough to produce sonorous continuity. Every schoolboy knows how to produce a note with his slate pencil. Holding the pencil vertically and somewhat loosely between the fingers, on moving it over the slate a succession of taps is heard. By pressure these taps can be caused to succeed each other so quickly as to produce a continuous sound. I will not call it musical, because this term is usually associated with pleasure, and the sound of the pencil is not pleasant. But it is not pleasant because it is not pure. It consists, in fact, of an assemblage of notes with plenty of discord among them.

The production of a musical sound by taps is usually effected by causing the teeth of a rotating wheel to strike in quick succession against a card. This was first illustrated by the celebrated Robert Hooke, to whom I have already referred,* and nearer our own day by the eminent French experimenter Savart. We will confine ourselves to

* On July 27, 1681, 'Mr. Hooke showed an experiment of making musical and other sounds by the help of teeth of brass wheels; which teeth were made of equal bigness for musical sounds, but of unequal for vocal sounds.' Birch's *History of the Royal Society*, p. 96, published in 1757.

The following extract is from the *Life of Hooke*, which precedes his *Posthumous Works*, published in 1705, by Richard Waller, Sec. R. S.:—'In July the same year he (Dr. Hooke) showed a way of making musical and other sounds by the striking of the teeth of several brass wheels, proportionally cut as to their numbers, and turned very fast round, in which it was observable that the equal or proportional stroaks of the teeth, that-is, 2 to 1, 4 to 3, &c., made the musical notes, but the unequal stroaks of the teeth more answered the sound of the voice in speaking.'

homelier modes of illustration. I have here a gyroscope, an instrument consisting mainly of a heavy brass ring *d*, fig. 13, loading the circumference of a disc, through which, and at right angles to its surface, passes a steel axis, delicately supported at its two ends. By coiling a string round the axis, and drawing it vigorously out, the ring is caused to spin rapidly; and along with it rotates a small toothed wheel W. I touch this wheel with the edge of a card *c*, and a musical sound of exceeding shrillness is the result. I place my thumb for a moment against the ring; the rapidity of its rotation is thereby diminished, and this is instantly announced by a lowering of the pitch of the note. By checking the motion still more, I lower the pitch still further. We are here made acquainted with the important fact that the pitch of a note depends upon the rapidity of its pulses.* At the end of the experiment you hear the separate taps of the teeth against the card, their succession not being quick enough to produce that continuous flow of sound which is the essence of music. A screw with a 'milled' head attached to a whirling table, and caused to rotate, produces by its taps against a card a note almost as clear and pure as that obtained from the toothed wheel of the gyroscope.

FIG. 13.

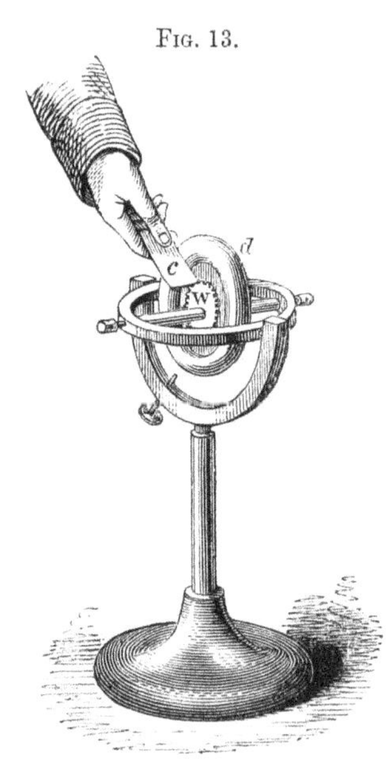

* Galileo finding the number of notches on his metal to be great when the pitch of the note was high, inferred that the pitch depended on the rapidity of the impulses.

The production of a musical sound by taps may also be pleasantly illustrated in the following way:—In this vice are fixed edgeways two pieces of sheet lead, with the edges a quarter of an inch apart. I lay this bar of brass across them, permitting it to rest upon the edges, and, tilting the bar a little with my hand, set it in oscillation like a see-saw. After a time, if left to itself, it comes to rest. But suppose the bar on touching the lead to be always tilted upwards by a force issuing from the lead itself, it is plain that the vibrations would then be rendered permanent. Now such a force is brought into play when the bar is heated. On its then touching the lead heat is

FIG. 14.

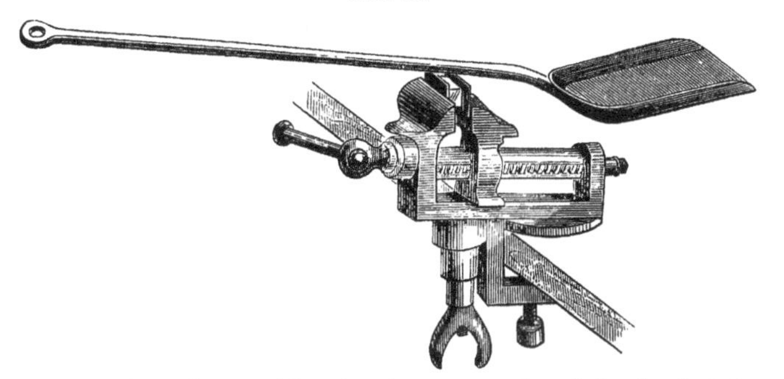

communicated, a sudden jutting upwards of the lead at the point of contact being the result. Hence an incessant tilting of the bar from side to side, so long as it continues sufficiently hot. Substituting for the brass bar the fire-shovel shown in fig. 14, the same effect is produced.

In its descent upon the lead the bar taps it gently, the taps in the case now before us being so slow that you may readily count them. But I have here a mass of metal differently shaped, which will vibrate more briskly and cause the taps to succeed each other with greater rapidity. I place this *rocker* upon a block of lead; the taps have hastened to a loud rattle audible to every person in this assembly. With the point of a file I press the rocker against the lead; the vibrations are thereby rendered more rapid, and the taps link themselves together to a deep musical tone. I have here another rocker which oscillates more quickly than the last one, and which produces music without any other pressure than that due to its own weight. Pressing it, however, with the file, I quicken its vibrations, the pitch rises, and now a note of singular force and purity fills the entire room. Relaxing the pressure, the pitch instantly falls; resuming the pressure, it again rises; and thus

Fig. 15.

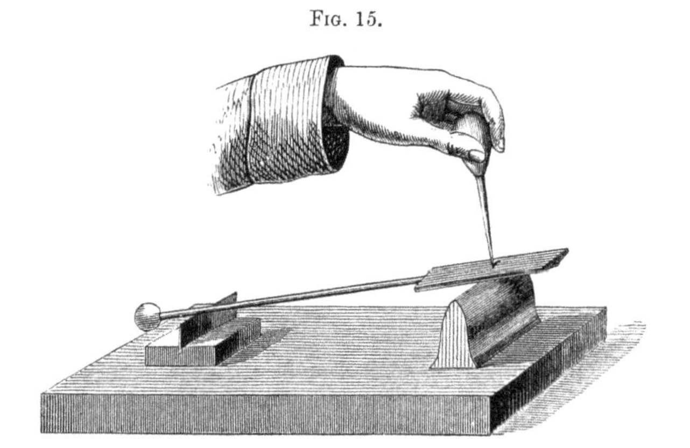

by the alteration of the pressure we obtain these amusing variations of tone. Finally, replacing this rocker by one of quicker vibration, we obtain a note of still higher pitch. The sound is here less easily obtained than with the last rocker, and it varies more capriciously. It is sometimes exceedingly shrill, subsiding at intervals into a kind of plaint, varying in its pitch from moment to moment. Instead of the full melodious note of the last rocker, we have here a kind of expostulatory screaming sound which resembles more than anything else the crying of an ill-tempered child. Nor are such rockers essential. Allowing one face of the clean square end of this heated poker to rest upon the block of lead, a rattle is heard; causing another face to rest upon the block, a clear musical note is obtained. The two faces have been bevelled differently by a file, so as to secure different rates of vibration.*

Professor Robison was the first to produce a musical sound by a quick succession of puffs of air. His device was the first form of an instrument which will soon be introduced to your notice under the name of the *syren*. Robison describes his experiment in the following words: —'A stopcock was so constructed that it opened and shut the passage of a pipe 720 times in a second. The apparatus was fitted to the pipe of a conduit leading from the bellows to the wind-chest of an organ. The air was simply allowed to pass gently along this pipe by the opening of the cock. When this was repeated 720 times in a second, the sound *g in alt* was most smoothly uttered, equal in sweetness to a clear female voice. When the frequency was reduced to 360, the sound was that of a clear, but rather a harsh man's voice. The cock was now altered in such a manner that it never shut the hole entirely, but left about one-third of it open. When this was repeated 720 times in a second, the sound was uncommonly smooth and sweet. When reduced to 360, the sound was more mellow than any man's voice of the same pitch.' I

* When a rough tide rolls in upon a pebbled beach, as at Blackgang Chine or Freshwater Gate in the Isle of Wight, the rounded stones are carried up the slope by the impetus of the water, and when the wave retreats the pebbles are dragged down. Innumerable collisions thus ensue of irregular intensity and recurrence. The union of these shocks impresses us as a kind of scream. Hence the line in Tennyson's *Maud*:—

'Now to the scream of a maddened beach dragged down by the wave.'

The height of the note depends in some measure upon the size of the pebbles, varying from a kind of roar—heard when the stones are large—to a scream; from a scream to a noise resembling that of frying bacon; and from this when the pebbles are so small as to approach the state of gravel, to a mere hiss. The roar of the breaking wave itself is mainly due to the explosion of bladders of air.

possess a cock fashioned like that of Professor Robison, and with it have verified his results.

But the difficulty of obtaining the necessary speed renders another form of the experiment more suited to our present purpose. Here is a disc of Bristol board B, fig. 16, twelve inches in diameter, and perforated at equal intervals along a circle near its circumference. The disc, being strengthened by a backing of tin, can be attached to a whirling table and caused to rotate rapidly. When it rotates the individual holes disappear, blending themselves in each case into a continuous shaded circle. Immediately over this circle is placed a glass tube *m*, which is connected

FIG. 16.

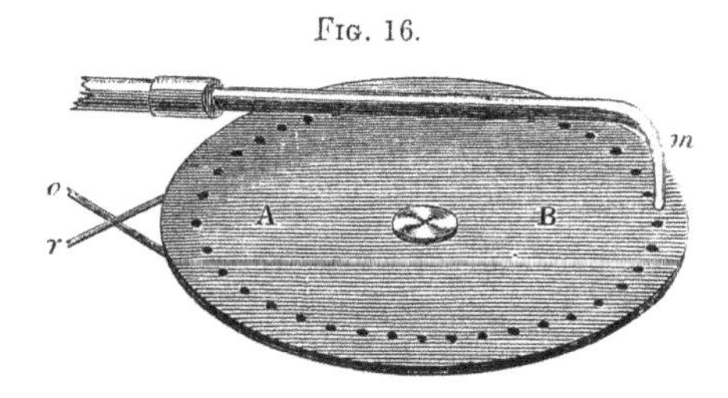

with a pair of acoustic bellows. The disc is now motionless, and the lower end of the glass tube is immediately over one of the perforations of the disc. If, therefore, the bellows be worked, the wind will pass from *m* through the hole underneath. But if the disc be turned a little, an unperforated portion of the disc comes under the tube, the current of air being then intercepted. As I thus turn the disc slowly I bring successive perforations under the tube, and whenever this occurs a puff of air gets through. The rotation is now rapid, and the puffs succeed each other in very quick succession, producing pulses in the air which blend to a continuous musical note, audible to you all. Mark how the note varies. When the whirling table is turned rapidly the sound is shrill; when its motion is slackened the pitch immediately falls. If instead of having a single glass tube I had two of them, the same distance apart as two of our orifices, so that whenever one tube stood over an orifice, the other should stand over the next one, it is plain that if both tubes were blown through, we should on turning the disc get a puff through two holes at the same time. The intensity of the sound would be thereby augmented, but the pitch would remain unchanged. The two puffs issuing at the same instant would act in concert, and produce a greater effect than one upon the ear. And if instead of two holes we had ten of them, or better still, if we had a tube for every orifice in the disc, the puffs from

Fig. 17.

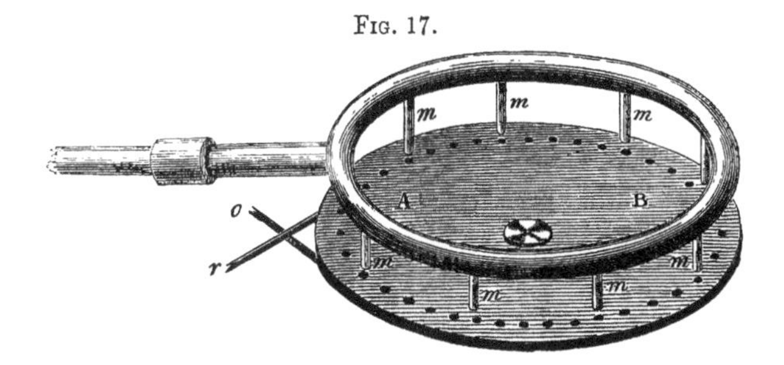

the entire series would all issue, and would all be cut off at the same time. These puffs would produce a note of far greater intensity than that obtained by the alternate escape and interruption of the air from a single tube. In the arrangement now before you, fig. 17, there are nine tubes through which the air is urged—through nine apertures, therefore, puffs escape at once. On turning the whirling table, and alternately increasing and relaxing its speed, the sound rises and falls like the wail of a changing wind.

Various other means may be employed to throw the air into a state of periodic motion. A stretched string pulled aside and suddenly liberated imparts vibrations to the air which succeed each other in perfectly regular intervals. A tuning-fork does the same. I draw this bow across the prongs of this tuning-fork, fig. 18. The bow is rubbed with resin, which enables the hairs to grip the fork. But the resistance of the fork soon becomes too strong, and the prong starts suddenly back; it is, however, immediately laid hold of again by the bow, to start back once more as soon as its resistance becomes great enough. This rhythmic process, continually repeated during the passage of the bow, finally throws the fork into a state of intense vibration, and the result is this clear full musical note. A person close at hand could

Fig. 18.

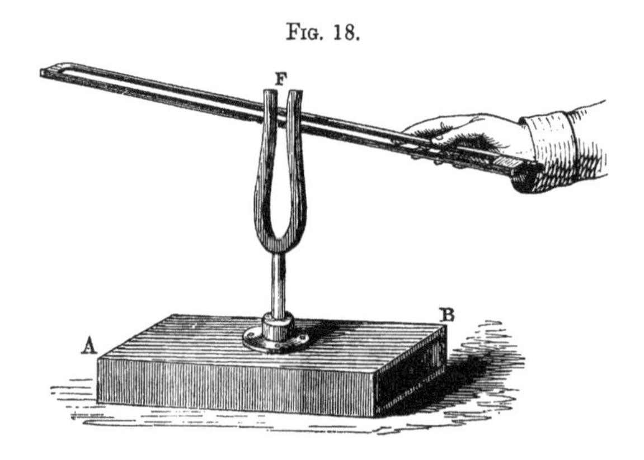

see the fork vibrating; a deaf person bringing his hand sufficiently near would feel the shivering of the air. I now cause its vibrating prong to touch a card, the taps against the card link themselves, as in the case of the gyroscope, to a musical sound, the fork coming rapidly to rest. It is now quite still, and what we call silence expresses this absence of motion.

When I first excite the tuning-fork the sound issues from it with maximum loudness, becoming gradually feebler as the fork continues to vibrate. I, being close to the fork, can notice at the same time that the amplitude or space through which the prongs oscillate becomes gradually less and less. But within the limits here employed the most expert ear in this assembly can detect no change in the pitch of the note. The lowering of the intensity of a note does not therefore imply the lowering of its pitch. In fact, though the amplitude changes, the rate of vibration remains the same. Pitch and intensity must therefore be held distinctly apart; the latter depends solely upon the amplitude, the former solely upon the rapidity of vibration.

This tuning-fork may be caused to write the story of its

Fig. 19.

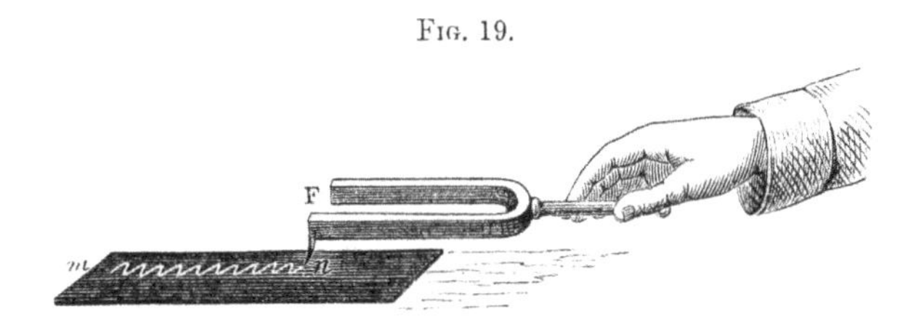

own motion; the mode of doing so being very easy of comprehension. Taking a piece of chalk in my fingers, and causing my hand to move up and down along this black board, I draw a short vertical line upon the board. As long as my hand is not permitted to move to the right or left, I simply go over the same chalked line. But if while the hand is thus moving up and down, I cause it also to move to the right or left, a sinuous line is drawn upon the board. The indentations indicate the vibration of my hand, and their depth indicates the amplitude of that vibration. Let us now turn to the tuning-fork. Attached to the side of one of its prongs F, fig. 19, is a thin strip of sheet copper which tapers to a point. I excite the tuning-fork; it vibrates, and the strip of metal accompanies it in its vibration. I bring the point of the strip gently down upon a piece of smoked glass. It moves to and fro over the smoked surface, leaving a clear line behind. As long as my hand is kept motionless, the point merely passes to and fro over the same line, just as the chalk a moment ago passed to and fro over the same line upon the black board. But it is plain that I have only to draw the fork along the glass to produce a sinuous line. I do so, and the line is drawn.

Placing the plate of glass in front of the electric lamp, I throw its magnified image upon a screen; the bright sinuous line is thus rendered visible to you all. While the plate is in front of the lamp, I will agitate the fork and draw it once more over the smoked surface. The luminous indented line starts instantly into existence. I repeat this process without exciting the fork afresh, and you observe how the depth of the indentations diminishes. The sinuous line approximates more and more to a straight one. This is the visual expression of decreasing amplitude. When the sinuosities entirely disappear the amplitude has become zero, and the sound which depends upon the amplitude ceases altogether.

To M. Lissajous we are indebted for a very beautiful method of giving optical expression to the vibrations of a tuning-fork. Attached to one of the prongs of this large fork F, fig. 20, is a small metallic mirror, the other prong being loaded with a piece of metal to establish equilibrium. Placing the fork in front of the electric lamp, and at a considerable distance from the screen, I permit a slender beam of intense light first to pass through a converging lens and then to fall upon the mirror. The beam is thrown back by reflection. In my hands I hold a small looking-glass, which receives the reflected beam, and from which it is again reflected to the screen. You see the image of the aperture through which the beam issues as a small luminous disc upon the white surface. It is perfectly motionless; but the moment the fork is set in vibration the reflected beam is tilted rapidly up and down, forming a band of light three feet long. The length of the band depends on the amplitude of the vibration, and you see it gradually shorten as the motion of the fork is expended. It remains, however, a straight line as long as the glass is held in a fixed position. But I now suddenly turn the glass so as to make the beam travel from left to right over the screen; and you observe the straight line instantly resolved into a beautiful luminous ripple *m n*. You are able to see this long wavy line because a luminous impression once made upon the retina lingers there for the tenth of a second. If then the

Fig. 20.

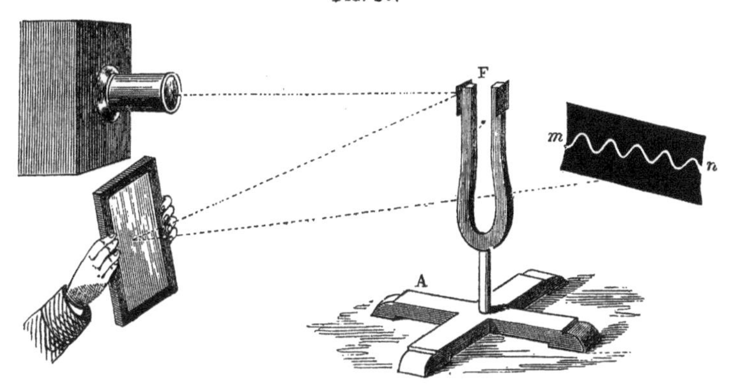

time required to transfer the elongated image from side to side of the screen be less than the tenth of a second, the wavy line of light will occupy for a moment the whole width of the screen. Instead of permitting the beam from the lamp to issue through a single aperture, I now cause it to issue through two apertures, about half an inch asunder, thus projecting two discs of light, one *above* the other, upon the screen. When the fork is excited and the mirror turned, we have this brilliant double sinuous line running over the dark surface, fig. 21. I now turn the diaphragm so as to place the two discs *beside* each other: on exciting the fork and moving the mirror at the proper rate, we obtain this beautiful figure produced by

Fig. 21.

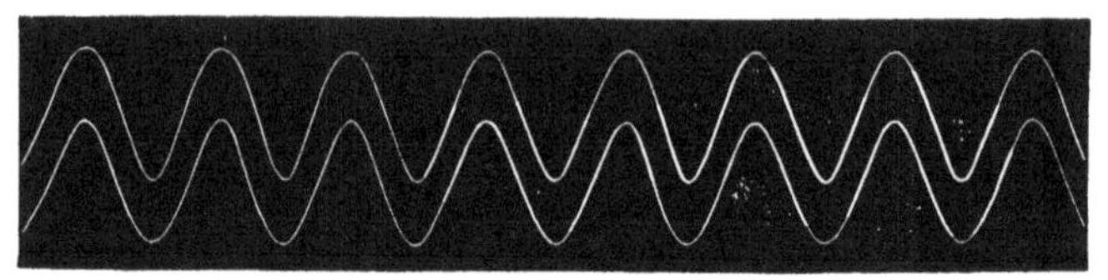

the interlacing of the two sinuous lines, fig. 22. The beauty of the figure, however, is not to be compared with the rippling of the actual lines of light over the screen.

Fig. 22.

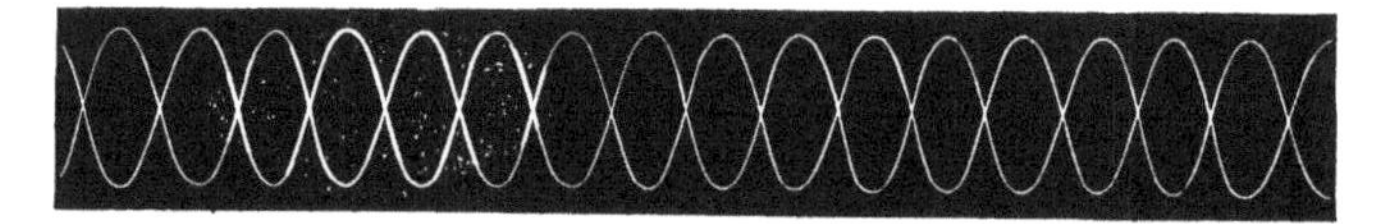

How are we to picture to ourselves the condition of the air through which this musical sound is passing? Imagine one of the prongs of the vibrating fork swiftly advancing; it compresses the air immediately in front of it, and when it retreats it leaves a partial vacuum behind, the process being repeated by every subsequent advance and retreat. The whole function of the tuning-fork is to carve the air into these condensations and rarefactions, and they, as they are formed, propagate themselves in succession through the air. A condensation with its associated rarefaction constitutes, as already stated, a sonorous wave. In water the length of a wave is measured from crest to crest; while in the case of sound the *wave-length* is given by the distance between two successive condensations. In fact, the condensation of the sound-wave corresponds to the crest, while the rarefaction of the sound-wave corresponds to the *sinus* of the water-wave. Let the dark spaces, *a*, *b*, *c*, *d*, fig. 23, represent the condensations, and the light ones, *a′*, *b′*, *c′*, *d′*, the rarefactions of the waves issuing from the fork A B: the wave-length would then be measured from *a* to *b*, from *b* to *c*, or from *c* to *d*.

FIG. 23.

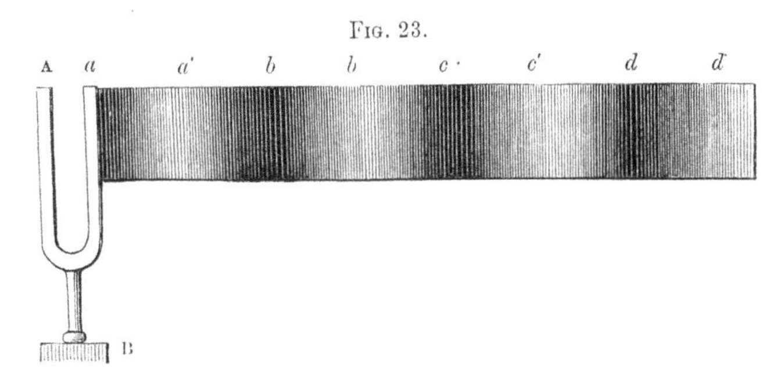

Pitch has been shown to depend upon rapidity of vibration. When two notes from two distinct sources are of the same pitch, their rates of vibration are the same. If, for example, a string yield the same note as a tuning-fork, it is because they vibrate with the same rapidity; and if a fork yield the same note as the pipe of an organ, or the tongue of a concertina, it is because the vibrations of the steel in the one case are executed in precisely the same time as the vibrations of the column of air, or of the tongue in the other. The same holds good for the human voice. If a string and a voice yield the same note, it is because the vocal chords of the singer vibrate in the same time as the string vibrates. Is there any way of determining the actual number of vibrations corresponding to a musical note? Can we infer from the pitch of a string, of an organ-pipe, of a tuning-fork, or of the human voice, the number of waves which it sends forth in a second? This very beautiful problem is capable of the most complete solution.

I have shown you, by the rotation of a perforated pasteboard disc, that a musical sound is produced by a quick succession of puffs. Had we any means of registering the number of revolutions accomplished by that disc in a minute, we should have in it a means of determining the number of puffs per minute due to a note of any determinate pitch. The disc, however, is but a cheap substitute for a far more perfect apparatus which I now bring before you; which requires no whirling table, and which registers its own rotations with the most perfect accuracy. I will take the instrument asunder, so that you may see its various parts. A brass tube *t*, fig. 24, leads into this brass cylinder C, closed at the top by a brass plate *a b*. This plate is perforated with four series of holes, placed along four concentric circles. The innermost series contains 8, the next 10, the next 12, and the outermost 16 orifices. When I blow into the tube *t*, the air escapes through the orifices, and the problem now before us is to convert these continuous currents into discontinuous puffs. This is accomplished by means of a brass disc, *d e*, also perforated with 8, 10, 12 and 16 holes, at the same distances from the centre and with the same intervals between them as those in the top of the cylinder C. Through the centre of the disc passes a steel axis, the two ends of which are

Fig. 24.

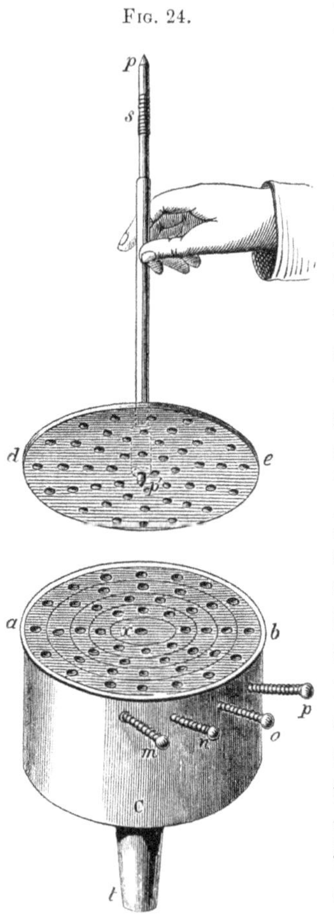

smoothly bevelled off to the points p and p'. My object now is to cause this perforated disc to rotate over the perforated top $a\ b$ of the cylinder C. You will understand how this is done by observing me put the instrument together. In the centre of $a\ b$, fig. 24, is a depression x sunk in steel, smoothly polished and intended to receive the end p' of the axis. I place the end p' in this depression, and holding the axis upright, bring down upon its upper end p a steel cap, finely polished within, which holds the axis at top and bottom, the pressure being so gentle, and the polish of the touching surfaces so smooth, that the disc can rotate with an exceedingly small amount of friction. At c, fig. 25, is the cap which fits on to the upper end of the axis $p\ p'$. In this figure the disc, $d\ e$, is shown covering the top of the cylinder C. I would ask you to neglect for the present the wheelwork of the figure. Turning the disc $d\ e$ thus slowly round, I can cause its perforations to coincide or not coincide with those of the cylinder underneath. As the disc turns, its orifices come alternately over the perforations of the cylinder, and over the spaces

Fig. 25.

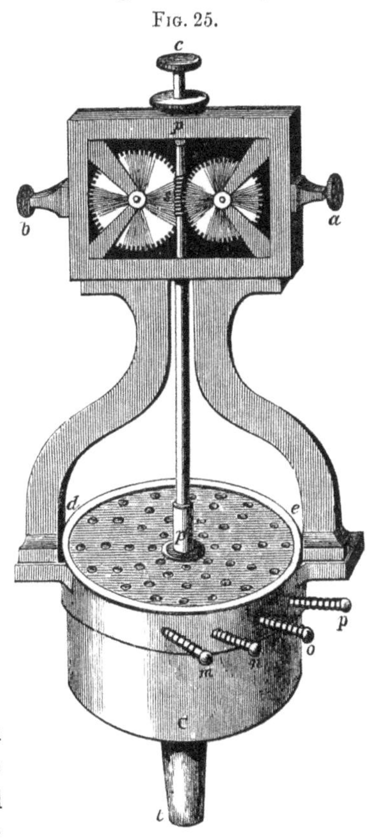

between the perforations. Hence it is plain that if air were urged into c, and if the disc could be caused to rotate at the same time, we should accomplish our object, and carve into puffs the streams of air. In this beautiful instrument, the disc is caused to rotate by the very air currents which it has to render intermittent. This is done by the simple device of causing the perforations to pass obliquely through the top of the cylinder c, and also obliquely, but oppositely inclined, through the rotating disc *d e*. The air is thus caused to issue from c, not vertically, but in side currents, which impinge against the disc and drive it round. In this way, by its passage through the syren, the air is moulded into sonorous waves.

Fig. 26.

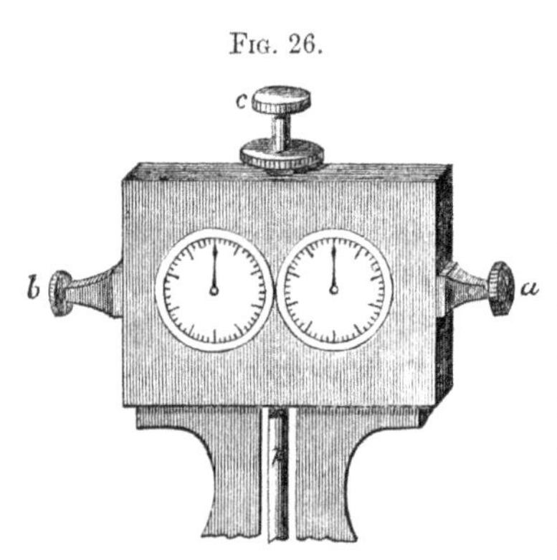

Another moment will make you acquainted with the recording portion of the instrument. At the upper part of the steel axis *p p′*, you observe a screw *s*, working into a pair of toothed wheels, seen when the back of the instrument is turned towards you, as in fig. 25. You notice that as the disc and its axis turn, these wheels rotate. In front you simply see these two graduated dials, fig. 26, each furnished with an index like the hand of a clock. These indexes record the number of revolutions executed by the disc in any given time. By pushing the button *a* or *b* it is in my power to throw this wheelwork into or out of action, and thus to start or to suspend, in a moment, the process of recording. Here, finally, is a series of pins, *m*, *n*, *o*, *p*, by which any series of orifices in the top of the cylinder c can be opened or closed at pleasure. By pressing *m*, I open one series; by pressing *n*, I open another. By pressing two keys, I open two series of orifices; by pressing three keys, I open three series; and by pressing all the keys, puffs issue from the four series simultaneously. The perfect instrument is now before you, and I think your knowledge of it is complete.

The instrument received the name of syren from its inventor, Cagniard de la Tour. The one now before you is the syren as greatly improved by Dove. The pasteboard syren, whose performance you have already heard, was devised by Seebeck, who gave the instrument various interesting forms, and executed with it various important experiments. Let us now make our syren sing. Pressing the key *m*, I open the outer series of apertures in the cylinder C, and, working the bellows, cause the air to impinge against the disc, which now begins to rotate. You hear a succession of puffs which follow each other so slowly that they may be counted. But as the motion augments, the puffs succeed each other with increasing rapidity, and now for the first time you hear a deep musical note. As the velocity of rotation increases the note rises in pitch: it is now very clear and full, and as I urge the air more vigorously, it becomes so shrill as to be painful to the ear. Here we have a further illustration of the dependence of pitch on rapidity of vibration. I touch the side of the disc and lower its speed; the pitch falls instantly. I continue the pressure, and the tone continues to sink, ending in the discontinuous puffs with which it commenced.

Were the blast sufficiently powerful, and the syren sufficiently free from friction, I might urge it to higher and higher notes, until finally its sound would become inaudible to human ears. This, however, would not prove the absence of vibratory motion in the air; but would rather show that our auditory apparatus is incompetent to take up, or our brains incompetent to translate into sound, vibrations whose rapidity exceeds a certain limit. The ear, as we shall immediately learn, is in this respect similar to the eye.

By means of this syren we can determine with extreme accuracy the rapidity of vibration of any sonorous body. It may be a vibrating string, an organ pipe, a reed, or the human voice. Operating delicately, we might even determine from the hum of an insect the number of times it flaps its wings in a second. I will illustrate the subject by determining in your presence this tuning-fork's rapidity of vibration. From the acoustic bellows I urge air through the syren, and, at the same time, draw my bow across the fork. Both now sound together, the tuning-fork yielding at present the highest note. But the pitch of the syren gradually rises, and now you hear the 'beats' so well known to musicians, which indicate that the two notes are not wide apart in pitch. These beats, you notice, become slower and slower; now they entirely vanish, and both notes blend as it were to a single stream of sound. The unison is now perfect, and by regulating the force of the bellows, I will endeavour to maintain it so. All this time the clockwork of the syren has remained out of action; I set it going by pushing the button *a*, as the second-hand of my watch crosses the number 60. I allow the disc to continue its rotations for a minute, exciting the tuning-fork from time to time to assure you and myself that the unison is preserved. The second-hand of my watch now approaches 60; as it passes the number I suddenly push *b* and stop the clock-work; and here recorded on the dials we have the exact number of revolutions performed by the disc. This number is 1,440. But the series of holes open during the experiment numbers 16; for every revolution, therefore, we had 16 puffs of air, or 16 waves of sound. Multiplying 1,440 by 16, we obtain 23,040 as the number of vibrations executed by the tuning-fork in a minute. Dividing this number by 60, we find the number of vibrations executed in a second to be 384.

Having determined the rapidity of vibration, the length of the corresponding sonorous wave is found with the utmost facility. Imagine this tuning-fork vibrating in free air. At the end of a second from the time it commenced its vibrations, the foremost wave would have reached a distance of 1,090 feet in air of the freezing temperature. In the air of this room, which has a temperature of about 15° C., it would reach a distance of 1,120 in a second. In this distance, therefore, are embraced 384 sonorous waves. Dividing, therefore, 1,120 by 384, we find the length of each wave to be nearly 3 feet. Determining in this way the rates of vibration of the four tuning-forks now before you, we find them to be 256, 320, 384, and 512; these numbers corresponding to wave-lengths of 4 feet 4 inches, 3 feet 6 inches, 2 feet 11 inches, and 2 feet 2 inches respectively. The waves generated by a man's organs of voice in common conversation are from 8 to 12 feet, those of a woman are from 2 to 4 feet in length. Hence a woman's ordinary pitch in the lower sounds of conversation is more than an octave above a man's; in the higher sounds it is two octaves.

And here it is important to note that when I speak of vibrations, I mean *complete ones*; and when I speak of a sonorous wave, I mean a condensation and its associated rarefaction. I include in one vibration one excursion *to and fro* of the vibrating body. Every wave generated by such vibrations bends the tympanic membrane once in and once out. These are the definitions of a vibration and of a sonorous wave employed in England and Germany. In France, however, a vibration consists of an excursion of the vibrating body *in one direction*, whether to or fro. The French vibrations, therefore, are only the halves of ours, and we therefore call them semi-vibrations. In all cases throughout these lectures, when the word vibration is employed without qualification, it refers to complete vibrations.

During the time required by each of those sonorous waves to pass entirely over a particle of air, that particle accomplishes one complete vibration. It is at one moment pushed forward into the condensation, while at the next moment it is urged back into the rarefaction. The time required by the particle to execute a complete oscillation is, therefore, that required by the sonorous wave to move through a distance equal to its own length. Supposing the length of the wave to be 8 feet, and the velocity of sound in air of our present temperature to be 1,120 feet a second, the wave in question will pass over its own length of air in $\frac{1}{140}$th of a second, and this is the time required by every air-particle that it passes to complete an oscillation. In air of a definite density and elasticity a certain length of wave always corresponds to the same pitch. But supposing the density or elasticity not to be uniform; supposing, for example, the sonorous waves from one of our tuning-forks to pass from cold to hot air, an instant augmentation of the wave-length would occur, without any change of pitch, for we should have no change in the rapidity with which the waves would reach the ear. Conversely with the same length of wave the pitch would be higher in hot air than in cold, for the succession of the waves would be quicker. In an atmosphere of hydrogen waves 8 feet long would produce a note nearly two octaves higher than in air, for, in consequence of the greater rapidity of propagation, the number of impulses received in a given time in the one case would be nearly four times the number received in the other.

I now open both the innermost and outermost series of the orifices of our syren. Sounding both of them, either together or in succession, the musical ears present at once detect the relationship of the two sounds. They notice at once that the sound which issues from the circle of 16 orifices is the octave of that which issues from the circle of 8. But for every wave sent forth by the latter, two waves are sent forth by the former. In this way we prove that the physical meaning of the term octave is that it is a note produced by double the number of vibrations of its fundamental. By multiplying the vibrations of the octave by two, we obtain *its* octave, and by a continued multiplication of this kind we obtain a series of numbers answering to a series of octaves. Starting, for example, from a fundamental note of 100 vibrations, we should find by this continual multiplication that a note five octaves above it would be produced by 3,200 vibrations. Thus:—

100	Fundamental note
2	
200	1st octave
2	
400	2nd octave
2	
800	3rd octave
2	
1600	4th octave
2	
3200	5th octave

The same figure is more readily obtained by multiplying the vibrations of the fundamental note by the fifth power of two. In a subsequent lecture we shall return to this question of musical intervals. For my present purpose it is only necessary that I should define an octave.

I have already stated that there is a limit to the ear's range of hearing. It is in fact limited in both directions in its perception of musical sounds. Savart fixed the lower limit of the human ear at eight complete vibrations a second; and to cause these slowly recurring vibrations to link themselves together, he was obliged to employ shocks of great power. By means of a toothed wheel and an associated counter, he fixed the upper limit of hearing at 24,000 vibrations a second. Helmholtz has recently fixed the lower limit at 16 vibrations, and the higher at 38,000 vibrations, a second. By employing very small tuning-forks, the late M. Depretz showed that a sound corresponding to 38,000 vibrations a second is audible.* Starting from the note 16 and multiplying continually by 2; or more compendiously raising 2 to the 11th power, and multiplying this by 16, we should find that at 11 octaves above the fundamental note the number of vibrations would be 32,768. Taking, therefore, the limits assigned by Helmholtz, the entire range of the human ear embraces about 11 octaves. But all the notes comprised within these limits cannot be employed in music. The practical range of musical sounds is comprised between 40 and 4,000 vibrations a second, which amounts, in round numbers, to 7 octaves.†

* The error of Savart consists, according to Helmholtz, in having adopted an arrangement in which overtones (described in Lecture III.) were mistaken for the fundamental one. The determination of the higher limit is worthy of still further experiment. Nor has it, as far as I can see, as yet been clearly established, how far an augmentation of intensity would affect the lower limit.

† 'The deepest tone of orchestra instruments is the E of the double-bass, with $41\frac{1}{4}$ vibrations. The new pianos and organs go generally as far as C^{I} with 33 vibrations; new grand pianos may reach A^{II} with $27\frac{1}{2}$ vibrations. In large organs a lower octave is introduced reaching to C^{II} with $16\frac{1}{2}$ vibrations. But the musical character of all these tones under E is imperfect, because they are near the limit where the power of the ear to unite the vibrations to a tone ceases. In height the pianoforte reaches to a^{iv} with 3,520 vibrations, or sometimes to c^{v} with 4,224 vibrations. The highest note of the orchestra is probably the d^{v} of the piccolo flute, with 4,752 vibrations.'—Helmholtz, *Tonempfindungen*, p. 30. In this notation we start from C with 66 vibrations, calling the first lower octave C^{I}, and the second C^{II}; and calling the first highest octave c, the second c^{I}, the third c^{II}, the fourth e^{I22}, &c. In England the deepest tone, Mr. Macfarren informs me, is not E but A, a fourth above it.

The limits of hearing are different in different persons. Dr. Wollaston, to whom we owe the first proof of this, while endeavouring to estimate the pitch of certain sharp sounds, remarked in a friend a total insensibility to the sound of a small organ-pipe, which, in respect to acuteness, was far within the ordinary limits of hearing. The sense of hearing of this person terminated at a note four octaves above the middle E of the pianoforte. The squeak of the bat, the sound of a cricket, even the chirrup of the common house-sparrow are unheard by some people who for lower sounds possess a sensitive ear. The ascent of a single note is sometimes sufficient to produce the change from sound to silence. 'The suddenness of the transition,' writes Wollaston, 'from perfect hearing to total want of perception, occasions a degree of surprise which renders an experiment of this kind with a series of small pipes among several persons rather amusing. It is curious to observe the change of feeling manifested by various individuals of the party, in succession, as the sounds approach and pass the limits of their hearing. Those who enjoy a temporary triumph are often compelled, in their turn, to acknowledge to how short a distance their little superiority extends.' 'Nothing can be more surprising,' writes Sir John Herschel, in reference to this subject, 'than to see two persons, neither of them deaf, the one complaining of the penetrating shrillness of a sound, while the other maintains there is no sound at all. Thus, while one person mentioned by Dr. Wollaston could but just hear a note 4 octaves above the middle E of the pianoforte, others have a distinct perception of sounds full 2 octaves higher. The chirrup of the sparrow is about the former limit; the cry of the bat about an octave above it; and that of some insects probably another octave.' In 'The Glaciers of the Alps' I have referred to a case of short auditory range noticed by myself, in crossing the Wengern Alp in company with a friend. The grass at each side of the path swarmed with insects, which to me rent the air with their shrill chirruping. My friend heard nothing of this, the insect-music lying quite beyond his limit of audition.

Behind the tympanic membrane exists a cavity—the drum of the ear—in part filled by a series of bones which cross it, and in part occupied by air. This cavity communicates with the mouth by means of a duct called the Eustachian tube. This tube is generally closed, the air-space behind the tympanic membrane being thus shut off from the external air. If, under these circumstances, the external air become denser, it will press the tympanic membrane inwards. If, on the other hand, the air outside become rarer, while the Eustachian tube remains closed, the membrane will be pressed outwards. Pain is felt in both cases, and partial deafness is experienced. Once while crossing the Stelvio Pass by night in company with a friend, he complained of acute pain in the ears. I desired him to swallow his saliva: he did so, and the pain instantly disappeared. By the act of swallowing the Eustachian tube is opened, and thus equilibrium is established between the external and internal pressure.

It is possible to quench the sense of hearing of low sounds by stopping the nose and mouth, and expanding the chest, as in the act of inspiration. This effort partially exhausts the space behind the tympanic membrane, which is then thrown into a state of tension by the pressure of the outward air. A similar deafness to low sounds is produced when the nose and mouth are stopped, and a strong effort is made to expire. In this case air is forced through the Eustachian tube into the drum of the ear, the tympanic membrane being distended by the pressure of the internal air. The experiment may be made in a railway carriage, when the low rumble will vanish, or be greatly enfeebled, while the sharper sounds are heard with undiminished intensity. Dr. Wollaston was expert in closing the Eustachian tube, and leaving the space behind the tympanic membrane occupied by either compressed or rarefied air. He was thus able to cause his deafness to continue for any required time without any effort on his part, always, however, abolishing it by the act of swallowing. A sudden concussion may produce deafness by forcing air either into or out of the drum of the ear. In the summer of 1858 I was on the Fee Alp in Switzerland, where, jumping from a cliff on to what I supposed to be a deep snow-drift, I came into rude collision with a rock which the snow barely covered. The sound of the wind, the rush of the glacier torrents, and all the other noises which a sunny day awakes upon the mountains, instantly ceased. I could hardly hear the sound of my guide's voice. This deafness continued for half-an-hour; at the end of which time a suitable act opened the Eustachian tube, and restored, with the quickness of magic, the innumerable murmurs which filled the air around me.

Light, like sound, is excited by pulses or waves; and lights of different colours, like sounds of different pitch, are excited by different rates of vibration. But in its width of perception the ear exceedingly transcends the eye; for while the former ranges over 11 octaves, but little more than a single octave is possible to the latter. The quickest vibrations which strike the eye, as light, have only about twice the rapidity of the slowest;* whereas the quickest vibrations which strike the ear, as a musical sound, have more than two thousand times the rapidity of the slowest.

Dove, as we have seen, extended the utility of the syren

* It is hardly necessary to remark that the quickest vibrations and shortest waves correspond to the extreme violet, while the slowest vibrations and longest waves correspond to the extreme red, of the spectrum.

of Cagniard de la Tour by providing it with four series of orifices instead of one. By doubling all its parts, Helmholtz has recently added vastly to the power of the instrument. The double syren, as it is called, is now before you, fig. 27. It is composed of two of Dove's syrens, C and C′, the upper one of which is turned upside down. You will recognise in the lower syren the instrument with which you are already acquainted. You see the disc with its perforations, and the four pins employed to open the orifices. The discs of the two syrens have a common axis, so that when one of them rotates the other rotates with it. As in the former case, the number of revolutions is recorded by clockwork (omitted in the figure). When air is urged through the tube *t′* the upper syren alone sounds; when urged through *t*, the lower one only sounds; when it is urged simultaneously through *t′* and *t*, both the syrens sound. With this instrument, therefore, we shall be able to introduce much more varied combinations than with the former one. Helmholtz has also contrived a means by which not only the disc of the upper syren, but the cylinder C′ above the disc can be caused to rotate. This is effected by a toothed wheel and pinion, turned by a handle. Underneath the handle is a dial with an index, the use of which will be subsequently illustrated.

Let us direct our attention for the present to the upper syren. By means of an india-rubber tube I connect the orifice *t′* with our acoustic bellows, and urge air into C′. Its disc turns round, and we obtain with it all the results already obtained with Dove's syren. The pitch of the note is now uniform. I turn the handle above, so as to cause the orifices of the cylinder C′ to *meet* those of the disc; it is plain that the two sets of apertures must now pass each other more rapidly than when the cylinder stood still. You perceive the result: an instant rise of pitch occurs when the handle is thus turned. By reversing the motion of the

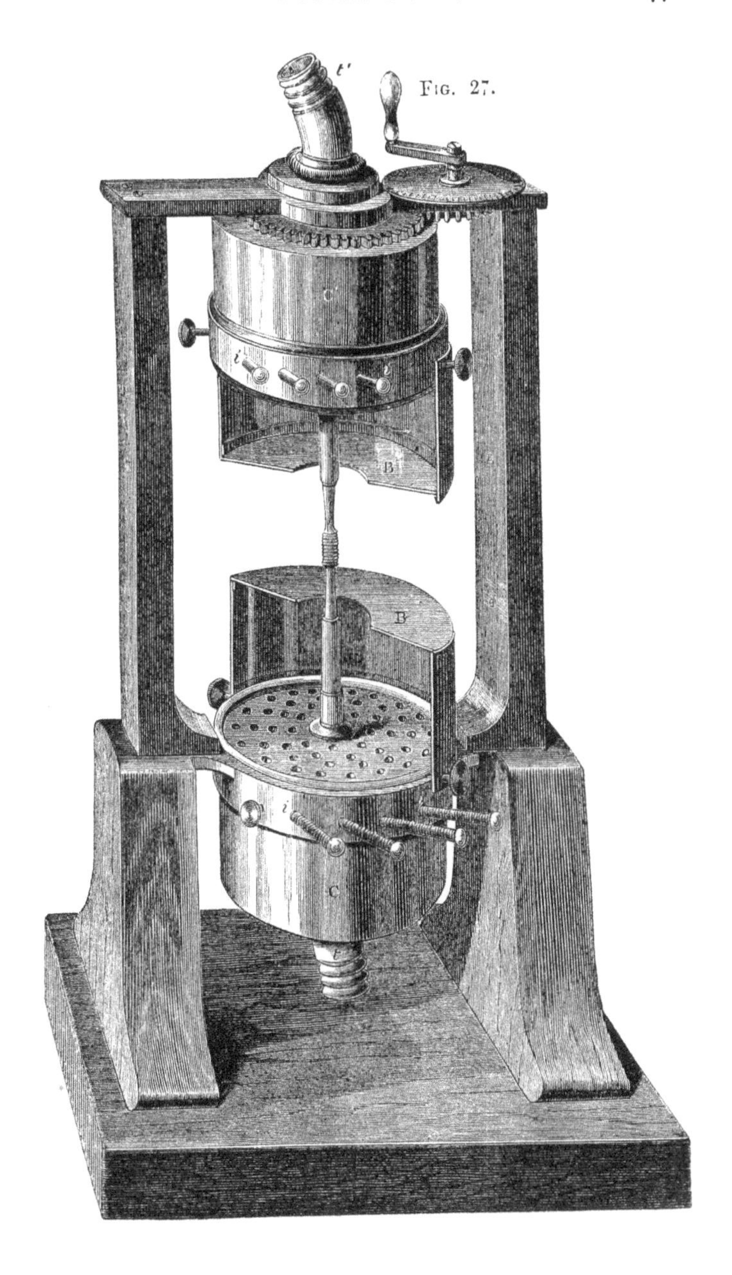

Fig. 27.

handle, the orifices are caused to pass each other more slowly than when the cylinder c′ is still, and in this case you notice an instant fall of pitch when the handle is turned. Thus, by imparting in quick alternation a right-handed and left-handed motion to the handle, we obtain these successive rises and falls of pitch. An extremely instructive effect of this kind may be observed at any railway station on the passage of a rapid train. During its approach the sonorous waves emitted by the whistle are virtually shortened, a greater number of them being crowded into the ear in a given time. During its retreat we have a virtual lengthening of the sonorous waves. The consequence is, that when approaching, the whistle sounds a higher note, and when retreating it sounds a lower note, than if the train were still. A fall of pitch, therefore, is perceived as the train passes the station.* This is the basis of Doppler's theory of the coloured stars. He supposes that all stars are white, but that some of them are rapidly retreating from us, thereby lengthening their luminiferous waves, and becoming red. Others are rapidly approaching us, thereby shortening their luminiferous waves, and becoming green or blue. The ingenuity of this theory is extreme, but its correctness is more than doubtful.

We have thus far occupied ourselves with the transmission of musical sounds through air. They are also faithfully transmitted by liquids and solids. To prove this, I place this drinking glass upon the table, and fill it with water. I strike this tuning-fork and cause it to vibrate; but except the persons closest to me, nobody present is conscious of its vibrations. The fork is screwed into a little wooden foot, which I now dip into the water, not permitting it to touch the sides of the glass. You instantly hear a

* Experiments on this subject were first made by M. Buys Ballot on the Dutch railway, and subsequently by Mr. Scott Russell in this country.

musical sound. I have here a tube M N, fig. 28, three feet long, set upright upon this wooden tray A B. The tube ends in a funnel at the top, and I now fill it with water to the brim. As before, I throw the fork F into vibration, and on dipping its foot into the funnel at the top of the tube, a musical sound swells out. I must so far forestall matters as to remark, that in this experiment the tray is the real sounding body. It has been thrown into vibration by the fork, but the vibrations have been conveyed to the tray *by the water.* Through the same medium vibrations are communicated to the auditory nerve, the terminal filaments of which are immersed in a liquid: substituting mercury for water, a similar result is obtained.

FIG. 28.

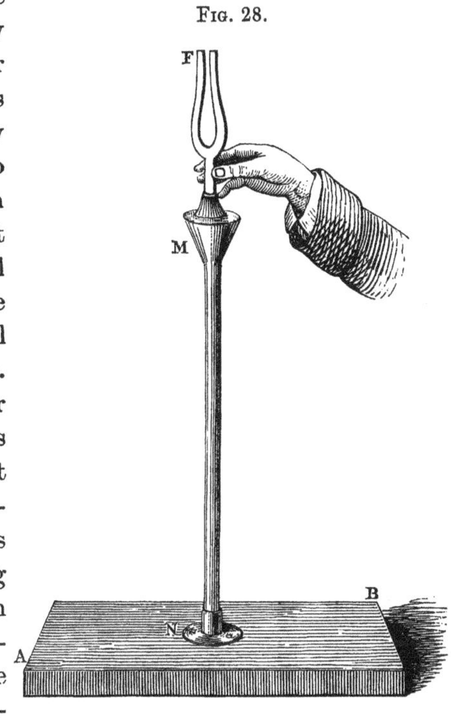

The syren has received its name from its capacity to sing under water. The vessel now in front of the table is half filled with water, in which a syren is wholly immersed. By turning this cock I permit the water from the pipes which supply the house to force itself through the instrument. Its disc is now rotating, and a sound of rapidly augmenting pitch issues from the vessel. The pitch rises thus rapidly because the heavy and powerfully pressed water soon drives the disc up to its maximum speed of rotation. I partially stop the supply; the motion relaxes and the pitch falls. Thus, by alternately opening and closing the cock, I cause the song of the syren to rise and fall in a wild and melancholy manner. You would not consider such a sound likely to woo mariners to their doom.

The transmission of musical sounds through solid bodies is also capable of easy and agreeable illustration. Before you is a wooden rod, thirty feet long, passing from this table through a window in the ceiling into the open air above. The lower end of the rod rests upon a wooden tray, to which I hope to be able to transfer the music of a body applied to the upper end of the rod. My assistant is above, with a tuning-fork in his hand. He strikes the fork against a pad; it vibrates, but you hear nothing. He now applies the stem of his fork to the end of the rod, and instantly the wooden tray upon the table is rendered musical. The pitch of this sound, moreover, is exactly that of the tuning-fork; the wood has been passive as regards pitch, transmitting the precise vibrations imparted to it, without any alteration. I employ another fork and obtain a note of another pitch. Thus I might employ fifty forks instead of two, and 300 feet of wood instead of 30; the wood would transmit the precise vibrations imparted to it, and no other.

We are now prepared to appreciate an extremely beautiful experiment, for which we are indebted to Professor Wheatstone, and which I am now able to make before you. In a room underneath this, and separated from it by two floors, is a piano. Through the two floors passes a tin tube 2½ inches in diameter, and along the axis of this tube passes a rod of deal, the end of which emerges from the floor in front of the lecture table. The rod is clasped by india-rubber bands, which entirely close the tin tube. The lower end of the rod rests upon the sound-board of the piano, its upper end being exposed before you. An artist is at this moment engaged at the instrument, but you hear no sound. I place this violin upon the end of the rod; the violin becomes instantly musical, not however with the vibrations of its own strings, but with those of the piano. I remove the violin, the sound ceases; I put in its place a guitar, and the music revives. For the violin and guitar I substitute this plain wooden tray; it is also rendered musical. Here, finally, is a harp, against the sound-board of which I cause the end of the deal rod to press; every note of the piano is reproduced before you. I lift the harp so as to break its connexion with the piano, the sound vanishes; but the moment I cause the sound-board to press upon the rod, the music is restored. The sound of the piano so far resembles that of the harp that it is hard to resist the impression that the music you hear is that of the latter instrument. An uneducated person might well believe that witchcraft is concerned in the production of this music.

What a curious transference of action is here presented to the mind! At the command of the musician's will, his fingers strike the keys; the hammers strike the strings, by which the rude mechanical shock is shivered into tremors. The vibrations are communicated to the sound-board of the piano. Upon that board rests the end of the deal rod, thinned off to a sharp edge to make it fit more easily between the wires. Through this edge, and afterwards along the rod, are poured with unfailing precision the entangled pulsations produced by the shocks of those ten agile fingers. To the sound-board of the harp before you the rod faithfully delivers up the vibrations of which it is the vehicle. This second sound-board transfers the motion to the air, carving it and chasing it into forms so transcendently complicated, that confusion alone could be anticipated from the shock and jostle of the sonorous waves. But the marvellous human ear accepts every feature of the motion; and all the strife and struggle and confusion melt finally into music upon the brain.*

* An ordinary musical box may be substituted for the piano in this experiment.

SUMMARY OF LECTURE II.

A musical sound is produced by sonorous shocks which follow each other at regular intervals, with a sufficient rapidity of succession.

Noise is produced by an irregular succession of sonorous shocks.

A musical sound may be produced by *taps* which rapidly and regularly succeed each other. The taps of a card against the cogs of a rotating wheel are usually employed to illustrate this point.

A musical sound may also be produced by a succession of *puffs*. The syren is an instrument by which such puffs are generated.

The pitch of a musical note depends solely on the number of vibrations concerned in its production. The more rapid the vibrations, the higher the pitch.

By means of the syren the rate of vibration of any sounding body may be determined. It is only necessary to render the sound of the syren and that of the body identical in pitch, to maintain both sounds in unison for a certain time, and to ascertain, by means of the counter of the syren, how many puffs have issued from the instrument in that time. This number expresses the number of vibrations executed by the sounding body.

When a body capable of emitting a musical sound—a tuning-fork for example—vibrates, it moulds the surrounding air into sonorous waves, each of which consists of a condensation and a rarefaction.

The length of the sonorous wave is measured from

condensation to condensation, or from rarefaction to rarefaction.

The wave-length is found by dividing the velocity of sound per second by the number of vibrations executed by the sounding body in a second.

Thus a tuning-fork which vibrates 256 times in a second produces in air of 15° C., where the velocity is 1,120 feet a second, waves 4 feet 4 inches long. While two other forks, vibrating respectively 320 and 384 times a second, generate waves 3 feet 6 inches and 2 feet 11 inches long.

A vibration, as defined in England and Germany, comprises a motion to *and* fro. It is a *complete* vibration. In France, on the contrary, a vibration comprises a movement to *or* fro. The French vibrations are with us semi-vibrations.

The time required by a particle of air over which a sonorous wave passes, to execute a complete vibration, is that required by the wave to move through a distance equal to its own length.

The higher the temperature of the air, the longer is the sonorous wave corresponding to any particular rate of vibration. Given the wave-length and the rate of vibration, we can readily deduce the temperature of the air.

The human ear is limited in its range of hearing musical sounds. If the vibrations number less than 16 a second, we are conscious only of the separate shocks. If they exceed 38,000 a second, the consciousness of sound ceases altogether. The range of the best ear covers about 11 octaves, but an auditory range limited to 6 or 7 octaves is not uncommon.

The sounds available in music are produced by vibrations comprised between the limits of 40 and 4,000 a second. They embrace 7 octaves.

The range of the ear far transcends that of the eye, which hardly exceeds an octave.

By means of the Eustachian tube, which is opened in the act of swallowing, the pressure of the air on both sides of the tympanic membrane is equalised.

By either condensing or rarefying the air behind the tympanic membrane, deafness to sounds of low pitch may be produced.

On the approach of a railway train the pitch of the whistle is higher, on the retreat of the train the pitch is lower, than if the train were at rest.

Musical sounds are transmitted by liquids and solids. Such sounds may be transferred from one room to another; from the ground-floor to the garret of a house of many stories, for example, the sound being unheard in the rooms intervening between both, and rendered audible only when the vibrations are communicated to a suitable sound-board.

LECTURE III.

VIBRATIONS OF STRINGS — HOW EMPLOYED IN MUSIC — INFLUENCE OF SOUND-BOARDS — LAWS OF VIBRATING STRINGS — ILLUSTRATIONS ON A LARGE SCALE—COMBINATION OF DIRECT AND REFLECTED PULSES—STATIONARY AND PROGRESSIVE WAVES—NODES AND VENTRAL SEGMENTS—APPLICATION OF RESULTS TO THE VIBRATIONS OF MUSICAL STRINGS—EXPERIMENTS OF M. MELDE—STRINGS SET IN VIBRATION BY TUNING-FORKS — LAWS OF VIBRATION THUS DEMONSTRATED — HARMONIC TONES OF STRINGS—DEFINITIONS OF TIMBRE OR QUALITY, OF OVERTONES AND CLANG—ABOLITION OF SPECIAL HARMONICS—CONDITIONS WHICH AFFECT THE INTENSITY OF THE HARMONIC TONES — OPTICAL EXAMINATION OF THE VIBRATIONS OF A PIANO-WIRE.

WE have to begin our studies to-day with the vibrations of strings, or wires; to learn how bodies of this form are rendered available as sources of musical sounds, and to investigate the laws of their vibrations.

Fig. 29.

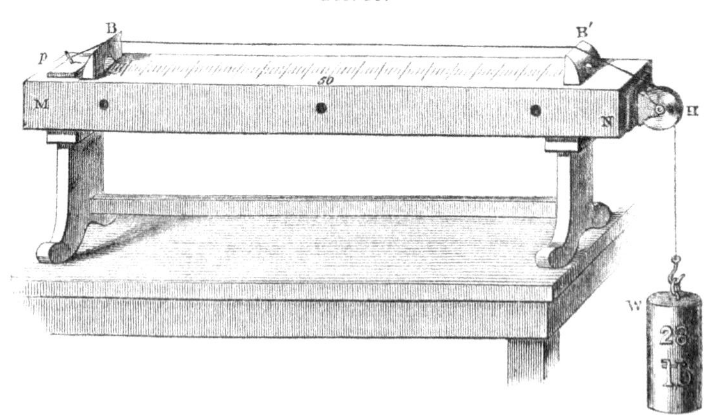

To enable a string to vibrate transversely, it must be stretched between two rigid points. Before you, fig 29, is an instrument employed to stretch strings, so as to render their vibrations audible. From the pin *p*, to which one end of it is firmly attached, the string passes across the two bridges B and B′, being afterwards carried over the wheel H, which moves with great freedom. The string is finally stretched by a weight W of 28 lbs. attached to its extremity. The bridges B and B′, which constitute the real ends of the string, are fastened on the long wooden box M N. The whole instrument is called a monochord or sonometer.

I take hold of the stretched string B B′ at its middle point, pull it aside, and liberate it suddenly. Let us henceforth call this act *plucking* the string. After having been plucked, the string springs back to its first position, passes it, returns, and thus vibrates for a time to and fro across its position of equilibrium. You hear a sound, and I at the same time can plainly see the limits between which the string vibrates. The sonorous waves which at present strike your ears do not proceed immediately from the string. The amount of motion which so thin a body imparts to the air is too small to be sensible at any distance. But the string is drawn tightly over the two bridges B B′; and when it vibrates, its tremors are communicated through these bridges to the entire mass of the box M N, and to the air within the box, which thus become the real sounding bodies.

That the vibrations of the string alone are not sufficient to produce the sound, may be thus experimentally demonstrated:—A B, fig. 30, is a piece of wood placed across an iron bracket C. From each end of the piece of wood depends a rope ending in a loop, while stretching across from loop to loop is an iron bar *m n*. From the middle of the iron bar hangs a steel wire *s s′*, stretched by a weight W of 28 lbs. By this arrangement, the wire is detached from all large surfaces to which it could impart its vibrations. Here is a second wire *t t′*, fig. 31, of the same length, thickness, and material as *s s'*, with one of its ends attached to the wooden tray A B. This wire also carries a weight W of 28 lbs. Finally, passing over the bridges B B' of the sonometer, fig. 29, is our third wire, in every respect like the two former, and like them stretched by a weight W of 28 lbs. I now pluck the wire *s s'*, fig. 30. It vibrates vigorously, but even those upon the nearest benches do not hear any sound. The agitation which it imparts to the air is too inconsiderable to affect the auditory nerve at any distance. I cause the wire *t t'*, fig. 31, to vibrate, and you all hear its sound distinctly. Though one end only of the wire is connected with the tray, A B, the vibrations transmitted to it are sufficient to convert the tray into a sounding body. Finally, I pluck the wire of the sonometer M N, fig. 29; here the sound is loud and full, because the instrument is specially constructed to take up the vibrations of the wire.

Fig. 30.

The importance of employing proper sounding apparatus in stringed instruments is rendered manifest by these experiments. It is not the chords of a harp, or a lute, or a piano, or a violin, that throw the air into sonorous vibrations. It is the large surfaces with which the strings are associated, and the air enclosed by these surfaces. The goodness of such instruments depends almost wholly upon the quality and disposition of their sound-boards.*

Fig. 31.

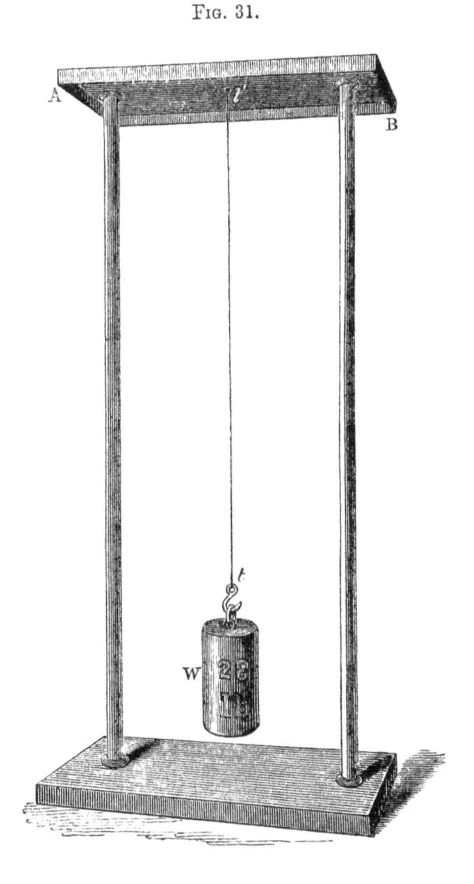

Take the violin as an example. It is, or ought to be, formed of wood of the most perfect elasticity. Imperfectly elastic wood expends the motion imparted to it in the friction of its own molecules; the motion is converted into heat instead of sound. The strings of the violin pass from the tail-piece of the instrument over the bridge, being thence carried to the pegs, the turning of which regulates the tension of the strings. The bow is drawn at a point about one-tenth of the length of the string from the bridge.

* To show the influence of a large vibrating surface in communicating sonorous motion to the air, Mr. Kilburn encloses a musical box within cases of thick felt. Through the cases a wooden rod, which rests upon the box, issues. When the box plays a tune, it is unheard as long as the rod only emerges; but when a thin disc of wood is fixed on the rod, the music becomes immediately audib e.

The two feet of the bridge rest upon the most yielding portion of the 'belly' of the violin, that is, the portion that lies between the two *f* shaped orifices. One foot is fixed over a short rod, the 'sound post,' which runs from belly to back through the interior of the violin. This foot of the bridge is thereby rendered rigid, and it is mainly through the other foot, which is not thus supported, that the vibrations are conveyed to the wood of the instrument, and thence to the air within and without. The molecular changes which age brings along with it are of importance. As the electrician finds that the glass he rubs to-day yields electricity not as it did a year ago; not, perhaps, as it will a year to come, the molecular condition of the glass, and with it its electric quality, changing through time; so do we find the sonorous quality of the wood of a violin mellowed by age. Moreover, the very act of playing has a beneficial influence; apparently constraining the molecules of the wood, which in the first instance were refractory, to conform at last to the requirements of the vibrating strings.

Having thus learned how the vibrations of strings are rendered available in music, we have next to investigate the laws of such vibrations. Laying hold of it at its middle point, I pluck the string B B′, fig. 29. The sound now heard is the fundamental or lowest note of the string, to produce which it swings, as a whole, to and fro. Placing a moveable bridge under the middle of the string, and pressing the string against the bridge, I divide it into two equal parts. Plucking either of those at its centre, a musical note is obtained, which many of you recognise as the octave of the fundamental note. Now, in all cases, and with all instruments, the octave of a note is produced by doubling the number of its vibrations. It can, moreover, be proved, both by theory and by the syren, that this half string vibrates with exactly twice the rapidity of the whole. In the same way it can be proved that one-third of the string vibrates with three times the rapidity, producing a note a fifth above the octave, while one-fourth of the string vibrates with four times the rapidity, producing the double octave of the whole string. In general terms, *the number of vibrations is inversely proportional to the length of the string.*

Again, the more tightly a string is stretched the more rapid is its vibration. I cause this comparatively slack string to vibrate, and you hear its low fundamental note. By turning a peg, round which one end of it is coiled, I tighten the string; the pitch is now higher. Taking hold with my left hand of the weight W, attached to the wire B B′ of our sonometer, and plucking the wire with the fingers of my right, I alternately press upon the weight and lift it. The quick variations of tension are expressed by this varying wailing tone. Now, the number of vibrations executed in the unit of time bears a definite relation to the stretching force. Applying different weights to the end of the wire B B′, and determining in each case the number of vibrations executed in a second, we find the numbers thus obtained to be *proportional to the square roots of the stretching weights.* A string, for example, stretched by a weight of 1 lb. executes a certain number of vibrations per second; if we wish to double this number, we must stretch it by a weight of 4 lbs.; if we wish to treble the number, we must apply a weight of 9 lbs., and so on.

The vibrations of a string also depend upon its thickness; preserving the stretching weight, the length, and the material of the string constant, *the number of vibrations varies inversely as the thickness of the string.* If, therefore, of two strings of the same material, equally long and equally stretched, the one has twice the diameter of the other, the thinner string will execute double the number of vibrations of its fellow in the same time. If one string be three times as thick as another, the latter will execute three times the number of vibrations, and so on.

Finally, the vibrations of a string depend upon the density of the matter of which it is composed. A platinum wire and an iron wire, for example, of the same length and thickness, stretched by the same weight, will not vibrate with the same rapidity. For while the specific gravity of iron, or in other words its density, is 7·8, that of platinum is 21·5. All other conditions remaining the same, *the number of vibrations is inversely proportional to the square root of the density of the string.* If the density of one string, therefore, be one-fourth that of another of the same length, thickness, and tension, it will execute its vibrations twice as rapidly; if its density be one-ninth that of the other, it will vibrate with three times the rapidity, and so on. The two last laws, taken together, may be expressed thus:—*The number of vibrations is inversely proportional to the square root of the weight of the string.*

In the violin and other stringed instruments we avail ourselves of thickness instead of length to obtain the deeper tones. In the piano we not only augment the thickness of the wires intended to produce the bass notes, but we load them by coiling round them an extraneous substance. They resemble horses heavily jockeyed, and move more slowly on account of the greater weight imposed upon the force of tension.

These, then, are the four laws which regulate the transverse vibrations of strings. We now turn to certain allied phenomena, which, though they involve mechanical considerations of a rather complicated kind, may be completely mastered by an average amount of attention. And they *must* be mastered if we would thoroughly comprehend the philosophy of stringed instruments.

From the ceiling *c*, fig. 32, of this room hangs an india-rubber tube 28 feet long. I have filled the tube with sand to render its motions slow and more easily followed by the eye. I take hold of its free end *a*, stretch the tube a little, and by properly timing my impulses cause it to swing to and fro as a whole, as shown in the figure. It has its definite time of vibration dependent on its length, weight, thickness and tension, and my impulses must synchronise with that time.

I stop the motion, and now by a sudden jerk I raise a hump upon the tube, which runs along it as a pulse towards its fixed end, where the hump reverses itself, and runs back to my hand. At the fixed end of the tube, according to the law of reflection, the pulse reversed both its position and the direction of its motion. Supposing *c*, fig. 33, to be the fixed end of the tube, and *a* the end held in the hand; if the pulse on reaching *c* have the position shown in (1), after reflection it will have the position shown in (2). The arrows mark the direction of progression. The time required for the pulse to pass from my hand to the fixed end and back, is exactly that required to accomplish one complete vibration of the tube as a whole. It is indeed the addition of such pulses which causes the tube to continue to vibrate as a whole.

FIG. 32.

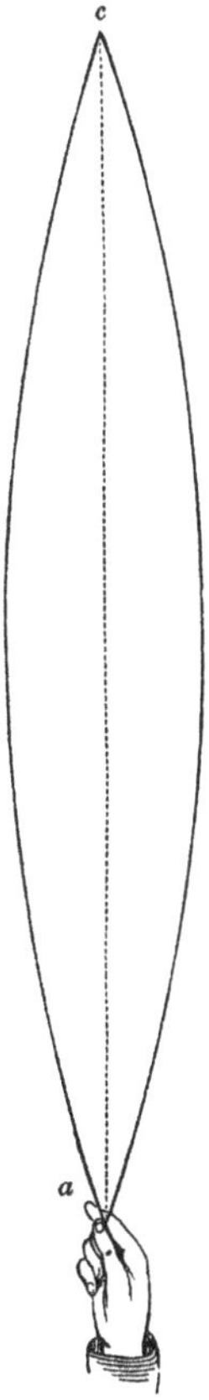

c

a

FIG. 33.

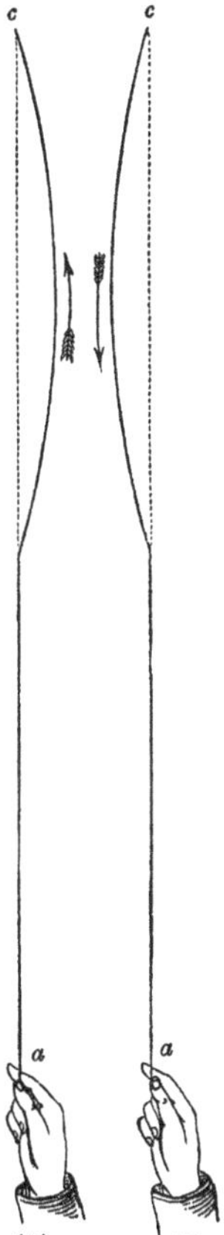

c *c*

a *a*

(1) (2)

If, instead of imparting a single jerk to the end of the tube, I impart a succession of jerks, thereby sending a series of pulses along the tube, every one of them will be reflected above, and we have now to enquire how the direct and reflected undulations behave towards each other.

I start a pulse along the tube. Let the time required for it to pass from my hand to the fixed end be one second; at the end of half a second it occupies the position *a b* (1), fig. 34, its foremost point having reached the middle of the tube. At the end of a whole second it would have the position *b c* (2), its foremost point having reached the fixed end *c* of the tube. At the moment when reflection begins at *c*, let another jerk be imparted at *a*; the reflected pulse from *c* moving with the same velocity as this direct one from *a*, the foremost points of both will arrive at the centre *b* (3) at the same moment. What must occur? The hump *a b* wishes to move on to *c*, and to do so must move the point *b* to the right. The hump *c b* wishes to move towards *a*, and to do so must move the point *b* to the left. The point *b*, urged by equal forces in two opposite directions at the same time, will not move in either direction. Under these circumstances, the two halves *a b*, *b c* of the tube will oscillate as if they were independent of each other (4). Thus by the combination of two *progressive pulses*, the one direct and the other reflected, we produce two *stationary pulses* on the tube *a c*.

Fig. 34.

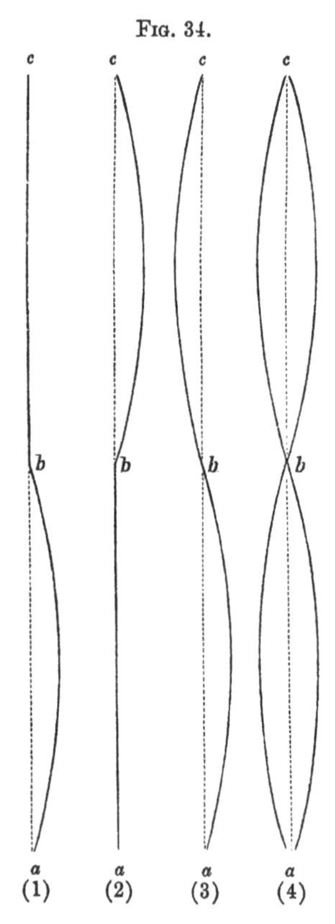

The vibrating parts *a b* and *b c* are called *ventral segments*; the point of no vibration *b* is called a *node*.

I use the term 'pulse' here advisedly, lest I should introduce confusion into your thoughts by the employment of the more usual term *wave*. For a wave embraces two of these pulses. It embraces both the hump and the depression which follows the hump. The length of a wave therefore, is twice that of a ventral segment.

FIG. 35.

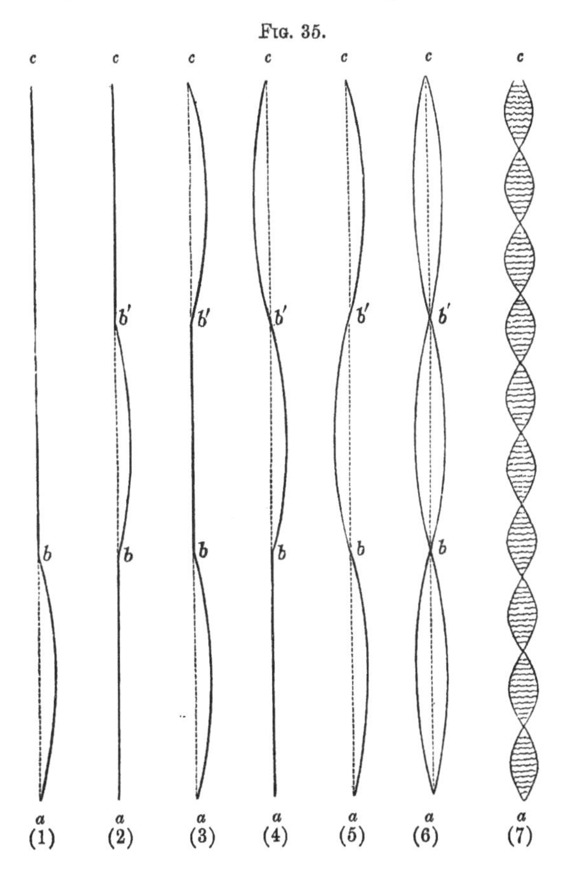

Supposing the jerks to be so timed as to cause each hump to be one-third of the tube's length. At the end of one-third of a second from starting the pulse will be in the position

$a\ b$ (1), fig. 35. In two-thirds of a second it will have reached the position $b\ b'$ (2), fig. 35. At this moment let a new pulse be started at a; after the lapse of an entire second from the commencement we shall have two humps upon the tube, one occupying the position $a\ b$ (3), the other the position $b'\ c$ (3). It is here manifest that the end of the reflected pulse from c, and the end of the direct one from a, will reach the point b' at the same moment. We shall therefore have the state of things represented in (4), where $b\ b'$ wishes to move upwards, and $c\ b'$ to move downwards. The action of both upon the point b' being in opposite directions, that point will remain fixed, *and from it, as if it were a fixed point, the pulse $b\ b'$ will appear to be reflected, while the segment $b'\ c$ will oscillate as an independent string.* Supposing that at the moment $b\ b'$ (4) begins to be reflected at b', we start another pulse from a, it will reach b (5) at the same moment the pulse reflected from b' reaches it. The pulses will neutralise each other at b, and we shall have there a second node. Thus, by properly timing our jerks, we divide the rope into three ventral segments, separated from each other by two nodal points. As long as the agitation continues the tube will vibrate as in (6).*

There is no theoretic limit to the number of nodes and ventral segments that may be thus produced. By quickening the impulses, I divide the tube into four ventral

* If, instead of moving the hand to and fro, it be caused to describe a small circle, the ventral segments become surfaces of revolution. Instead of the hand we may employ a hook turned by a wheel, or whirling table, and a string of catgut 10 or 12 feet long, with silvered beads strung along it, as a vibrating chord. Attaching one end of the string to the hook, the other end to a freely moving swivel connected with a fixed stand, on turning the wheel and properly regulating both the tension and the rapidity of rotation, the beaded chord may be caused to rotate as a whole, and to divide itself successively into 2, 3, 4, or 5 ventral segments. When the whole chord is enveloped in a cylinder of light from the electric lamp every bead describes a brilliant circle, and a very splendid experiment is the result.

segments separated by three nodes; here again I have five ventral segments and four nodes. With this particular tube the hand may be caused to vibrate sufficiently quickly to produce ten ventral segments, as shown in fig. 35 (7). When the stretching force is constant, the number of ventral segments is proportional to the rapidity of the hand's vibration. To produce 2, 3, 4, 10 ventral segments, requires twice, three times, four times, ten times the rapidity of vibration necessary to make the tube swing as a whole. When the vibration is very rapid, the ventral segments appear like a series of shadowy spindles, separated from each other by dark motionless nodes. The experiment is a beautiful one, and easily performed.

It is quite plain, that any other oscillating body whose vibrations are of sufficient power, and whose periods are of the proper kind, may be substituted for the hand. Fixing, for example, one end of a tolerably heavy rod in a vice, or, better still, screwing one end into an anvil or other heavy block, and attaching to the free end of the rod the end of our india-rubber tube; by varying the length of the rod its vibrations may be caused to synchronise with the various vibrations of the tube, and thus the latter may be caused to swing as a whole, or to divide itself into any number of oscillating parts.

The subject of stationary waves was first experimentally treated by the Messrs. Weber, in their excellent researches on Wave-motion. It is a subject which will well repay your attention by rendering many of the most difficult phenomena of musical strings perfectly intelligible to you. To make the connexion of both classes of vibrations more obvious, I will vary our last experiments. Before you is a piece of india-rubber tubing, 10 or 12 feet long, stretched from *a* to *c*, fig. 36, and made fast to two pins at *c* and *a*. The tube is blackened, and behind it is placed a surface of white paper, to render its motions more visible. I first encircle the tube at its centre b (1) by the thumb and forefinger of my left hand, and taking the middle of the lower half $b\ a$ of the tube in my right, I pluck it aside. Not

FIG. 36.

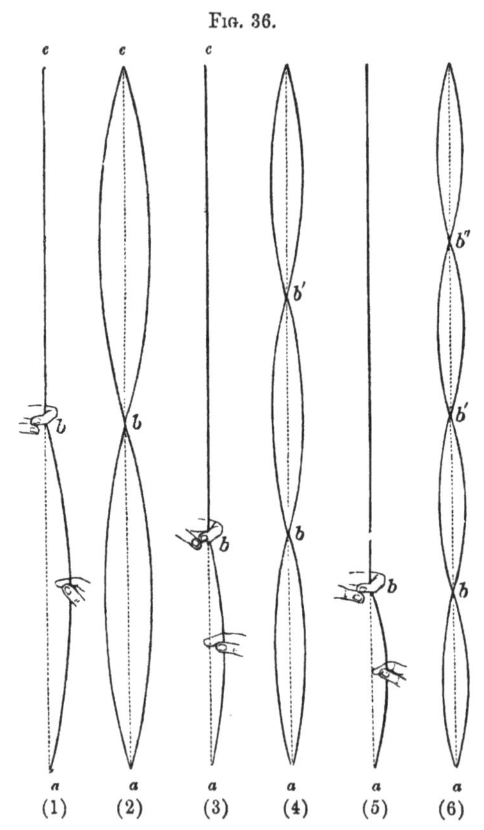

only does the lower half swing, but the upper half also is thrown into vibration. Withdrawing my hands wholly from the tube, its two halves $a\ b$ and $b\ c$ continue to vibrate, being separated from each other by a node at the centre (2).

I now encircle the tube at a point b (3) one-third of its length from its lower end a, and taking hold of $a\ b$ at its centre, I pluck it aside; the length *b c* above my hand instantly divides into two vibrating segments. Withdrawing my hands wholly, you see the entire tube divided into three ventral segments, separated from each other by two motionless nodes *b b′* (4). I pass on to the point *b* (5), which marks off one-fourth of the length of the tube, encircle it, and pluck the shorter segment aside. The longer segment above my hand divides itself immediately into three vibrating parts. So that, on withdrawing my hand, the whole tube appears before you divided into four ventral segments, separated from each other by three nodes *b b′ b″* (6). In precisely the same way I divide the tube into five vibrating segments with four nodes.

This sudden division of the long upper segment of the tube, without any apparent cause, is very surprising; but if you grant me your attention for a moment, you will find that these experiments are essentially similar to those which illustrated the coalescence of direct and reflected undulations. Reverting for a moment to the latter experiments, you observed that the to-and-fro motion of my hand through the space of a single inch, was sufficient to make the middle points of the ventral segments vibrate through the space of a foot or eighteen inches. By being properly timed the impulses accumulated, until the amplitude of the vibrating segments exceeded immensely that of the hand which produced them. My hand, in fact, constituted a nodal point, so small was its comparative motion. Indeed, it is usual, and correct, to regard the ends of the tube also as nodal points.

Consider now the case represented in (1) fig. 36, where the tube was encircled at its middle, the lower segment *a b* being thrown into the vibration corresponding to its length and tension. The circle formed by my finger and thumb permitted the tube to oscillate at the point *b* through the space of an inch; and the vibrations at that point acted upon the upper half *b c* exactly as my hand acted when it caused the tube suspended from the ceiling to swing as a whole, as in fig. 32. Instead of the timed vibrations of my hand, we have now the timed vibrations of the lower half of the tube; and these, though narrowed to an inch at the place clasped by my finger and thumb, soon accumulate, and finally produce an amplitude far exceeding their own. The same reasoning applies to all the other cases of subdivision. If instead of encircling a point by the finger and thumb and plucking the portion of the tube below it aside, that same point were taken hold of by the hand and agitated in the period proper to the lower segment of the tube, precisely the same effect would be produced. We thus reduce both effects to one and the same cause; namely, the combination of direct and reflected undulations.

And here, let me add, that when I divided the tube by the timed impulses of my hand, not one of its nodes was, strictly speaking, a point of no motion; for were the nodes not capable of vibrating through a very small amplitude, the motion of the various segments of the tube could not be maintained.

What is true of the undulations of an india-rubber tube applies to all undulations whatsoever. Water-waves, for example, obey the same laws, and by the coalescence of direct and reflected waves exhibit similar phenomena. I have here a long and narrow vessel with glass sides: a copy, in fact, of the wave canal of the brothers Weber. It is filled to the level A B, fig. 37, with coloured water. By tilting the end A, suddenly, I generate a wave, which moves on to B, and is there reflected. By sending forth fresh waves at the proper intervals, I divide the surface into two stationary undulations. Making the succession of impulses more rapid I can subdivide the surface into three, four (shown in the figure), or more stationary undulations, separated from each other by nodes. The step of a water-carrier is sometimes so timed as to throw the surface of the water in his vessel into stationary waves, which may augment in height until the water splashes over the brim. Practice has taught the water-carrier what to do; he changes his step, alters the period of his impulses, and thus stops the accumulation of the motion.*

Fig. 37.

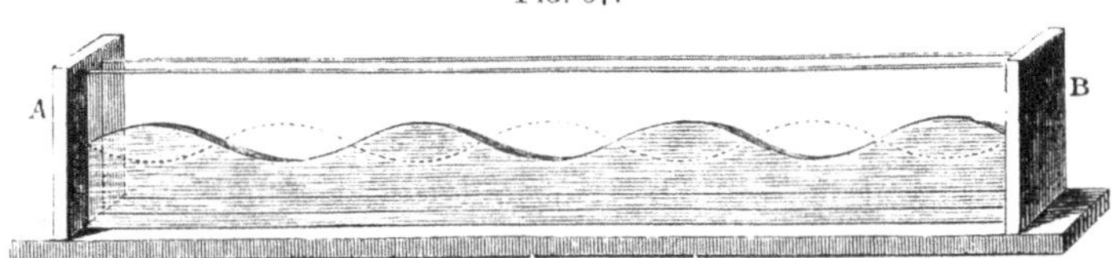

Let me now pass on from these gross, but not unbeautiful mechanical vibrations, to those of a sounding string. Here is our monochord with its steel wire, various lengths of which have been already sounded before you. In those experiments when I wished to shorten the wire I employed a movable bridge, pressing the wire against the bridge so as to deprive the point pressed of all possibility of motion. This strong pressure, however, is not necessary. I place the feather end of a goose-quill lightly against

* In travelling recently in the coupé of a French railway carriage I had occasion to place a bottle half filled with water on one of the little coupé tables. It was interesting to observe it. At times it would be quite still; at times it would oscillate violently. To the passenger within the carriage there was no change in the motion to which the difference could be ascribed. But the fact is that in the one case the tremor of the carriage contained no vibrations synchronous with the oscillating period of the water, while in the second case such vibrations were present. Out of the confused assemblage of tremors the water selected the particular constituent which belonged to itself, and declared its presence when the traveller was utterly unconscious of its introduction.

In a subsequent lecture I shall have occasion to speak of the dancing of flames in the carriages of the Metropolitan Railway in response to synchronous vibrations.

the middle of the string, and then draw a violin bow over one of its halves; the string yields the octave of the note yielded by the whole string. The mere damping of the string at the centre, by the light touch of the feather, is sufficient to cause the string to divide into two vibrating segments. Nor is it necessary that I should hold the feather there throughout the experiment: after having drawn the bow, I may remove the feather; the string will continue to vibrate, emitting the same note as before. We have here a case exactly analogous to that in which I damped the central point of our stretched india-rubber tube, by encircling it with my finger and thumb as in fig. 36 (1). Not only did the half which I plucked aside vibrate, but the upper half vibrated also. We can, in fact, reproduce, with the vibrating string, every effect obtained with the tube. This, however, is a point of such importance that I must not neglect to illustrate it experimentally.

To prove to you that when I damp the centre, and draw my bow across one of the halves of the string, the other half vibrates, I place across the middle of the

FIG. 38.

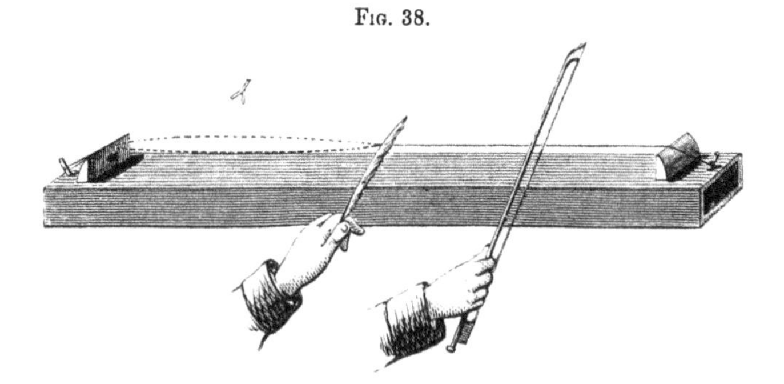

untouched half a little rider of red paper. I damp the centre and draw the bow, the string shivers, and the rider is overthrown, fig. 38.

I now damp the string at a point which cuts off one-third of its length, and then draw the bow across the shorter section. Not only is this section thereby thrown into vibration, but the longer section divides itself into two ventral segments with a node between them. I

FIG. 39.

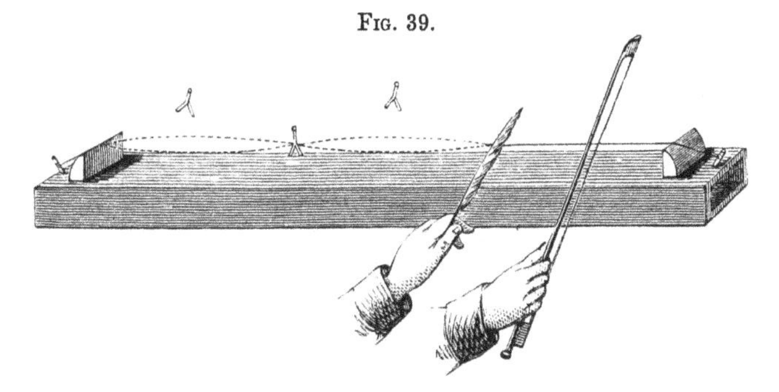

prove this by placing small riders of red paper on the ventral segments, and a rider of blue paper at the node. Passing the bow across the short segment you observe a fluttering of the red riders, and now they are completely tossed off, while the blue rider which crosses the node is undisturbed, fig. 39.

Again, I damp the string at the end of one-fourth of its length, and now affirm that when the bow is drawn

FIG. 40.

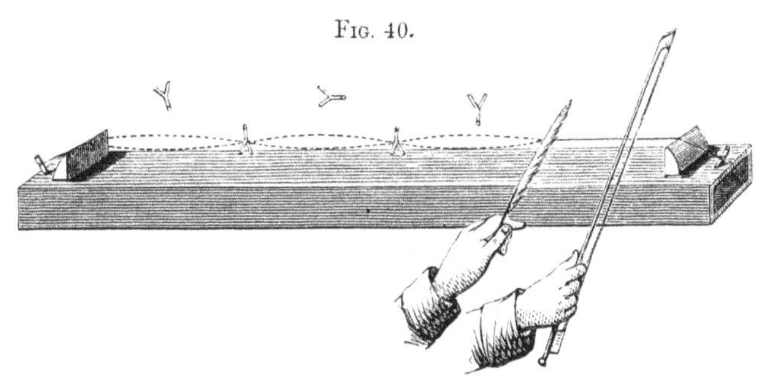

across the shorter section, the remaining three-fourths will divide themselves into three ventral segments, with two nodes between them. I prove my assertion by unhorsing the three riders placed astride the ventral segments, while the two at the nodes keep their places undisturbed, fig. 40.

Finally, I damp the string at the end of one-fifth of its length. Arranging, as before, the red riders on the ventral segments and the blue ones on the nodes; by a single sweep of the bow I unhorse the four red riders, and leave the three blue ones undisturbed, fig. 41. In this way we perform with a sounding string the same series of experiments that were formerly executed with a

FIG. 41.

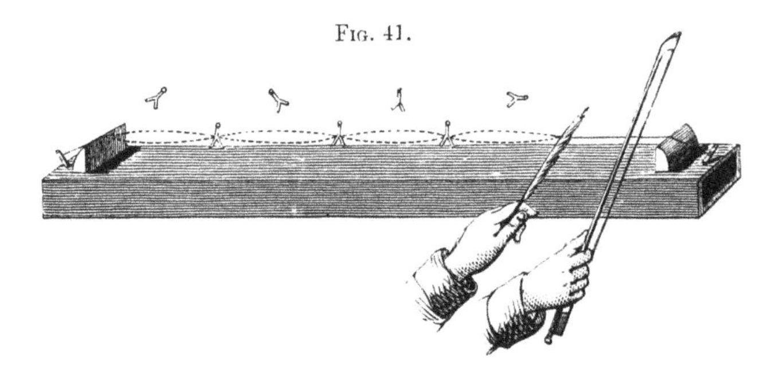

stretched india-rubber tube, the results in both cases being identical.*

To make this identity still more evident to you, I have stretched behind the table from side to side of the room a stout steel wire 28 feet in length. I take the central point of this wire between my finger and thumb, and allow my assistant to pull one-half of it aside, and then suddenly liberate it. It vibrates, and the vibrations transmitted to

* Chladni remarks (*Akustik*, p. 55) that it is usual to ascribe to Sauveur the discovery, in 1701, of the modes of vibration corresponding to the higher tones of strings; but that Noble and Pigott had made the discovery in Oxford in 1676, and that Sauveur declined the honour of the discovery when he found that others had made the observation before him.

the other half are sufficiently powerful to toss into the air a large sheet of paper placed astride the wire. With this long wire, and with riders not of one-eighth of a square inch, but of 30, 40, or 50 square inches in area, I repeat all the experiments which you have witnessed with the musical string. The sheets of paper placed across the nodes remain always in their places, while those placed astride the ventral segments are tossed simultaneously into the air when the shorter segment of the wire is set in vibration. In this case, when close to it, you can actually see the division of the wire.

It will not fail to interest you if I now introduce to your notice some recent experiments with vibrating strings, which appeal to the eye with a beauty and a delicacy far surpassing anything attainable with our sonometer. To M. Melde, of Marburg, we are indebted for this new method of exhibiting the vibrations of strings. I will perform his experiments on such a scale, and with such variations and modifications as our circumstances here suggest to us.

First, then, observe that I have here a large tuning-fork, T, fig. 42, with a small screw fixed into the top of one of its

Fig. 42.

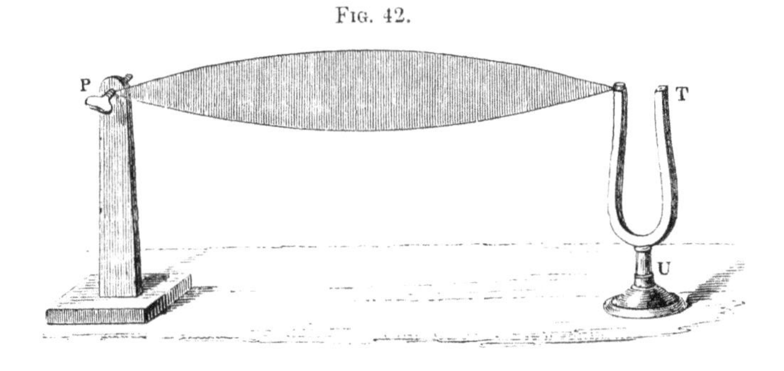

prongs, by which this silk string can be firmly attached to the prong. From the fork the string passes round a distant peg P, by turning which it may be stretched to any required extent. I draw my bow across the fork; an irregular flutter of the string is the only result. But I now tighten it, and at length, by adjusting the tension, cause it to expand into a beautiful gauzy spindle, more than six inches across at its widest part, and shining with a kind of pearly lustre. The stretching force at the present moment is such that the string swings to and fro as a whole, its vibrations being executed in a vertical plane.

I now relax the string, and when the proper tension has been reached, it suddenly divides into two ventral segments, separated from each other by a sharply-defined and apparently motionless node.

While the fork continues vibrating, I relax it still further; the string now divides into three vibrating parts. Slackening it still more, it divides into four vibrating parts. And thus I might continue to subdivide the string into ten, or even twenty ventral segments, separated from each other by the appropriate number of nodes.

When white silk strings vibrate thus, their beauty is extreme. The nodes appear perfectly fixed, while the ventral segments form spindles so delicate, that they seem woven from opalescent air; every protuberance of the twisted string, moreover, writes its motion in a more or less luminous line on the surface of the aërial gauze. The four modes of vibration just illustrated are represented in fig. 43, 1, 2, 3, 4.*

* The first experiment really made in the lecture was with a bar of steel 62 inches long, 1½ inch wide, and ½ an inch thick, bent into the shape of a tuning-fork, with its prongs 2 inches apart, and supported on a heavy stand. The chord attached to it was 9 feet long and a quarter of an inch thick. The prongs were thrown into vibration by striking them briskly with two pieces of lead covered with pads and held one in each hand. The prongs vibrated transversely to the chord. The vibrations produced by a single stroke were sufficient to carry the chord through several of its subdivisions and back to a single ventral segment. That is to say, by striking

When the synchronism is perfect, the vibrations of the string are steady and long-continued. A slight departure from synchronism, however, introduces unsteadiness, and the ventral segments, though they may show themselves for a time, quickly disappear.

FIG. 43.

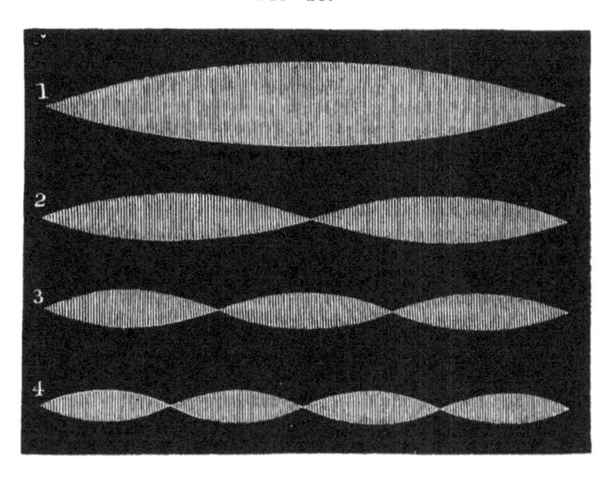

In the arrangement now before you, fig. 42, the fork vibrates in the direction of the length of the string. Every forward stroke of the fork raises a protuberance, which runs to the fixed end of the string, and is there reflected; so that when the *longitudinal* impulses are properly timed, they produce a *transverse* vibration. For the sake of illustration, I attach one end of a heavy chord or chain

the prongs and causing the chord to vibrate as a whole, it could, by relaxing the tension, be caused to divide into two, three, or four vibrating segments; and then, by increasing the tension, to pass back through four, three, and two divisions, to one, *without renewing the agitation of the prongs.* The chord was of such a character, that instead of oscillating to and fro in the same plane, each of its points described a circle. The ventral segments, therefore, instead of being flat surfaces were surfaces of revolution, and were equally well seen from all parts of the room. The tuning-forks employed in the subsequent illustrations were prepared for me by that excellent acoustic mechanician König, of Paris, being such as are usually employed in the projection of Lissajou's experiments.

to a hook, A, fig. 44, fixed in the wall, and laying hold of the other end with my hand, stretch the chord horizontally. I then move my hand to and fro in the direction of the chord. It swings as a whole, and you may notice that always, when the chord is at the limit of its swing, my hand is in its most forward position. If it vibrate in a vertical plane, my hand, in order to time the impulses properly, must be at its forward limit at the moment the chord reaches the upper boundary, and also at the moment it reaches the lower boundary of its excursion. A little reflection will make it plain that, in order to accomplish this, my hand must execute a complete vibration while the chord executes a semi-vibration; in other words, the vibrations of my hand must be twice as rapid as those of the chord.

FIG. 44.

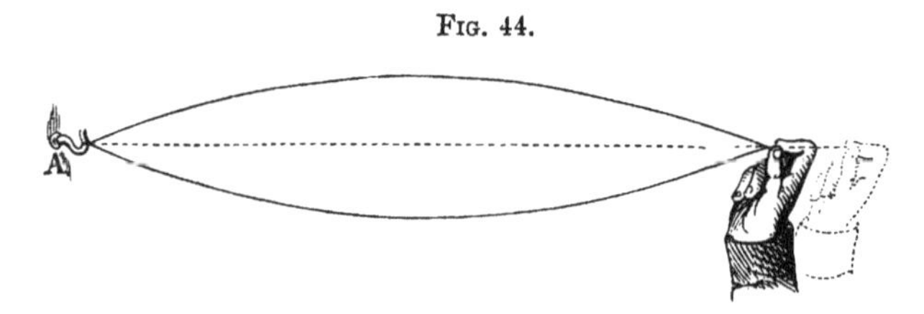

Precisely the same is true of our tuning-fork. When the fork vibrates in the direction of the string, the number of complete vibrations which it executes in a certain time is twice the number executed by the string itself. And if, while arranged thus, a fork and string vibrate with sufficient rapidity to produce musical notes, the note of the fork will be an octave above that of the string.

But if instead of moving my hand to and fro in the direction of this heavy chord, I move it at right angles to that direction, then every upward movement of my hand coincides with an upward movement of the chord; every downward movement of my hand with a downward movement of the chord. In fact, the vibrations of hand and string, in this case, synchronise perfectly; and if the hand could emit a musical note, the chord would emit a note of the same pitch. The same holds good when a vibrating fork is substituted for the vibrating hand.

Hence, if the string vibrate as a whole when the vibrations of the fork are along it, it will divide into two ventral segments when the vibrations are across it; or, more generally expressed, preserving the tension constant, whatever be the number of ventral segments produced by the fork when its vibrations are in the direction of the string, twice that number will be produced when the vibrations are transverse to the string. Here, for example, is a string A B, figs. 45 and 46, passing over a pulley B, and stretched by a definite weight (not shown in the figure). When the tuning-fork vibrates *along* it, as in fig. 45, the string

Fig. 45.

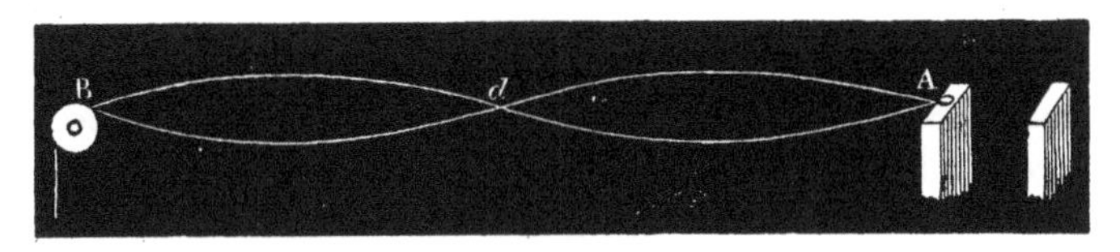

divides into two equal ventral segments. I turn the fork so that it shall vibrate at right angles to the string. The number of ventral segments is now four, fig. 46, or double the former number. Attaching two strings of the same

Fig. 46.

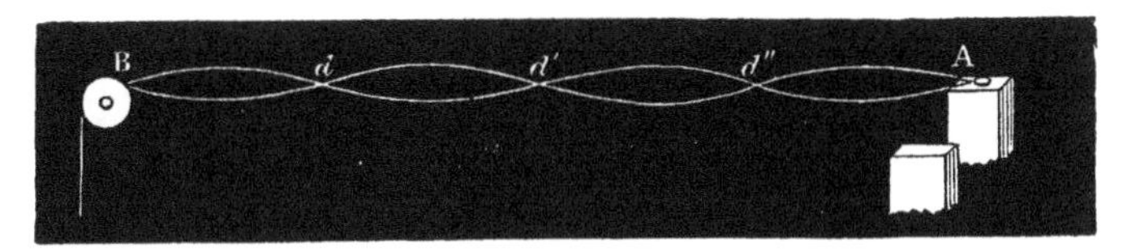

length to the same fork, the one parallel and the other perpendicular to the direction of vibration, and stretching both with equal weights; when the fork is caused to vibrate, one of them divides itself into twice the number of ventral segments exhibited by the other.

A number of exquisite effects may be obtained with these vibrating chords. The path described by any point of any one of them may be studied, after the manner of Dr. Young, by illuminating that point, and watching the line of light which it describes. This is well illustrated by a flat burnished silver wire, twisted so as to form a spiral surface, from which, at regular intervals, the light flashes when the wire is illuminated. Attached to a proper tuning-fork, and illuminated by the electric light, such a wire now steadily vibrates before you, its luminous spots describing straight lines of sunlike brilliancy. I now slacken the wire, but not so much as to produce its next higher subdivision. Upon the general motion of the wire we have now superposed a host of minor motions, the combination of all producing scrolls of marvellous complication and of indescribable splendour.*

In reflecting on the best means of rendering these beautiful effects visible to a large audience, the thought occurred to me of employing a fine platinum wire heated to redness by an electric current. Such a wire now stretches from this tuning-fork over a bridge of copper, and then passes round a peg. The copper bridge on the one hand, and the tuning-fork on the other, are the poles of a voltaic battery, from which a current passes through the wire and causes it to glow. I draw my bow across the fork; the wire vibrates as a whole: its two ends are brilliant, while its middle is dark, being chilled by its rapid passage through the air. Thus you have a shading off of incandescence from the ends to the centre of the

* I make no attempt to represent these beautiful effects by figures. The rapid rippling of the scrolls from one form of beauty to another cannot be rendered.

wire. I relax the tension, the wire now divides itself into two ventral segments; I relax still further, and obtain three; still further, and now you have the wire divided into four ventral segments, separated from each other by these three brilliant nodes. Right and left from every node the incandescence shades away until it disappears. You notice also when the wire settles into steady vibration, that the nodes shine out with greater brilliancy than did the wire before the vibration commenced. The reason is this. Electricity passes more freely along a cold wire than along a hot one. When, therefore, the vibrating segments are chilled by their swift passage through the air, their conductivity is improved, more electricity passes through the vibrating than through the motionless wire, and hence the augmented glow of the nodes. If, previous to the agitation of the fork, the wire be at a bright red heat, when it vibrates its nodes are raised to the temperature of fusion.

I shall now rapidly extend the experiments of M. Melde to the establishment of all the laws of vibrating strings. I have here four tuning-forks, *a*, *b*, *c*, *d*, whose rates of vibration are to each other as the numbers 1, 2, 4, 8. Attaching a string to the largest fork, *a*, I stretch it by a weight, which causes it to vibrate as a whole. Keeping the stretching weight the same, I determine the lengths of the same string, which, when attached to the other three forks, *b*, *c*, *d*, swing as a whole. The lengths in the four respective cases are as the numbers 8, 4, 2, 1.

From this follows the first law of vibration, which we have already established in another way, viz.:—*the length*

A string steeped in a solution of the sulphate of quinine, and illuminated by the violet rays of the electric lamp, exhibits brilliant fluorescence. When the fork to which it is attached vibrates, the string divides itself into a series of spindles, and separated from each other by more intensely luminous nodes, emitting a light of the most delicate greenish-blue.

of the string is inversely proportional to the rapidity of vibration.

In this case the longest string vibrates as a whole when attached to the fork *a*. I now transfer it to *b*, still keeping it stretched by the same weight. It vibrates when *b* vibrates; but how? By dividing into two equal ventral segments. In this way alone can it accommodate itself to the swifter vibrating period of *b*. Attached to *c*, the same string separates into four, while when attached to *d* it divides into eight ventral segments. The number of the ventral segments is proportional to the rapidity of vibration. It is evident that we have here, in a more delicate form, a result which we have already established in the case of our india-rubber tube set in motion by the hand. It is also plain that this result might be deduced theoretically from our first law.

We may extend the experiment. Here are two tuning-forks separated from each other by the musical interval called a fifth. Attaching a string to one of the forks, I stretch the string until it divides into two ventral segments; attached to the other fork, and stretched with the same weight, it divides instantly into three segments when the fork is set in vibration. Now to form the interval of a fifth, the vibrations of the one fork must be to those of the other in the ratio of 2 : 3. The division of the string, therefore, declares the interval. Here also are two forks separated by an interval of a fourth. With a certain tension one of the forks divides our string into three ventral segments; with the same tension the other fork divides it into four, which two numbers express the ratio of the vibrations. In the same way the division of the string in relation to all other musical intervals may be illustrated.*

* The subject of musical intervals will be treated in a subsequent lecture.

Again. Here are two tuning-forks, *a* and *b*, one of which (*a*) vibrates twice as rapidly as the other. I attach this string of silk to *a*, stretching the string until it synchronises with the fork, and vibrates as a whole. I now form a second string of the same length, by laying four strands of the first side by side. I attach this compound thread to *b*, and keeping the tension the same as in the last experiment, set *b* in vibration. The compound thread synchronises with *b*, and swings as a whole. Hence, as the fork *b* vibrates with half the rapidity of *a*, by quadrupling the weight of the string, I halved its rapidity of vibration. In the same simple way it might be proved that by augmenting the weight of the string nine times we reduce the number of its vibrations to one-third. We thus demonstrate the law:—

The rapidity of vibration is inversely proportional to the square root of the weight of the string.

A beautiful confirmation of this result is thus obtained: —Attached to this tuning-fork is a silk string six feet long. Two feet of the string are composed of four strands of the single thread, placed side by side, the remaining four feet are a single thread. I apply a stretching force, which causes the string to divide into two ventral segments. But how does it divide? Not at its centre, as is the case when the string is of uniform thickness throughout, but at the point where the thick string terminates. This thick segment, two feet long, is now vibrating at the same rate as the thin segment four feet long, a result which follows by direct deduction from the two laws already established. The result therefore reacts as a corroboration of these laws. I need hardly say that if the lengths were in any other ratio than 1 : 2, the node would not be formed at the point of junction of the two strings.

Here again are two strings of the same length and

thickness. One of them is attached to the fork *b*, the other to the fork *a*, which vibrates with twice the rapidity of *b*. Stretched by a weight of 20 grains, the string attached to *b* vibrates as a whole. Substituting the fork *a* for *b*, a weight of 80 grains causes the string to vibrate as a whole. Hence to double the rapidity of vibration, we must quadruple the stretching weight. In the same way it might be proved, that to treble the rapidity of vibration, we should have to make the stretching weight nine-fold. Hence our third law :—

The rapidity of vibration is proportional to the square root of the tension.

Let me vary this experiment. I carry this silk chord from the tuning-fork over the pulley, and stretch it by a weight of 80 grains. The string vibrates as a whole as at A, fig. 47. By diminishing the weight I relax the string, which finally divides sharply into two ventral segments, as at B. What is now the stretching weight? 20 grains, or one-fourth of the first. With a stretching weight of almost exactly 9 grains it divides into three segments, as at C; while with a stretching weight of 5 grains it divides into four segments, as at D. Thus, then, a tension of one-fourth doubles, a tension of one-ninth trebles, and a tension of one-sixteenth quadruples the number of ventral segments. In general terms, the number of segments is inversely proportional to the square root of the tension. This result may be deduced by reasoning from our first and third laws, and its realisation here confirms their correctness.

Finally, here are three wires of the same length and thickness, but of very different densities; one of them is of the light metal aluminium, another of silver, and another of the heavy metal platinum. I attach the wires in succession to this tuning-fork, and determine the weights necessary to cause them to vibrate as a whole, or to form the same number of ventral segments. The specific gravities, or densities, of the wires I find to be directly proportional to the stretching weights. But, other things being equal, the rapidity of vibration has been proved to be inversely proportional to the square roots of the stretching weights. Hence follows our fourth law, viz.:—

FIG. 47.

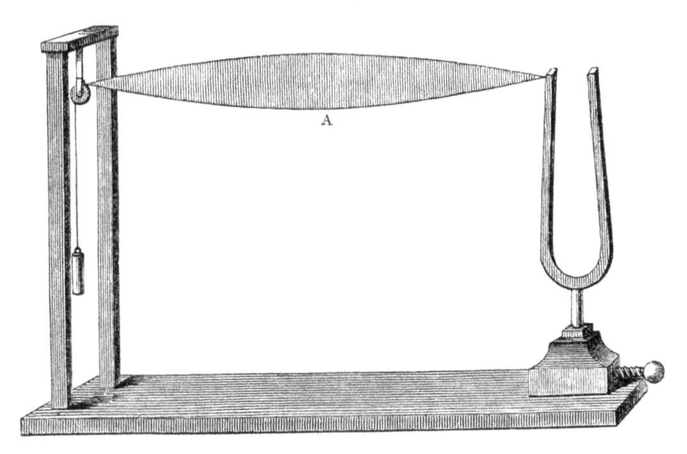

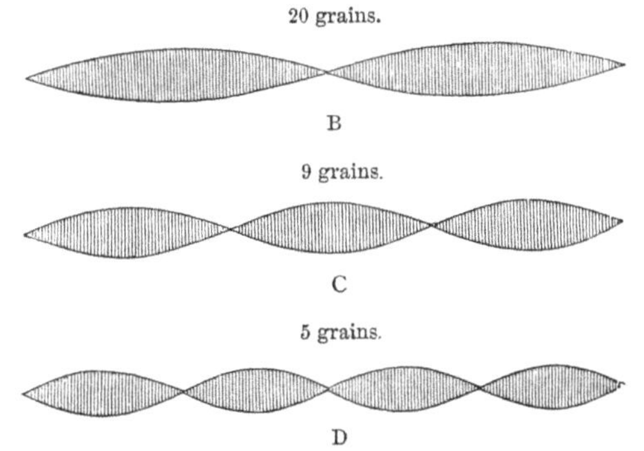

The rapidity of vibration of different wires of the same length and thickness is inversely proportional to the square roots of their densities.

It is evident that by means of a tuning-fork the

specific gravities of all metals capable of being drawn into wires of sufficient fineness and tenacity may be determined.*

Thus, by a series of reasonings and experiments totally different from those formerly employed, we arrive at the self-same laws. In science different lines of reasoning often abut upon the same truth; and if we only follow them faithfully, we are sure to reach that truth at last. We may emerge, and often do emerge, from our reasoning with a contradiction in our hands, but on retracing our steps we infallibly find the cause of the contradiction to be due not to any lack of constancy in nature, but of accuracy in man. It is the millions of experiences of this kind which science furnishes that give us our present faith in the stability of nature.

HARMONIC SOUNDS OR OVERTONES.

We now approach a portion of our subject which will subsequently prove to be of the very highest importance. It has been shown by the most varied experiments that a stretched string can either vibrate as a whole, or divide itself into a number of equal parts, each of which vibrates as an independent string. Now it is not possible to sound the string as a whole without at the same time causing, to a greater or less extent, its subdivision; that is to say, superposed upon the vibrations of the whole string we have always, in a greater or less degree, the vibrations of its aliquot parts. The higher notes produced by these latter vibrations are called the *harmonics* of the

* By the choice of suitable strings and tuning-forks, all the effects here described may be rendered visible to an audience of a thousand persons. The visibility is heightened by illuminating the vibrating strings with the electric light. They may be either looked at directly, in which way their beauty is best seen, or their shadows may be cast upon a white surface. In this way the division of thin short strings may be rendered visible.

string. And so it is with other sounding bodies; we have in all cases a coexistence of vibrations. Higher tones mingle with the fundamental one, and it is their intermixture which determines what, for want of a better term, we call the *quality* of the sound. The French call it *timbre,* and the Germans call it *Klangfarbe.** It is this union of high and low tones that enables us to distinguish one musical instrument from another. A clarionet and a violin, for example, though tuned to the same fundamental note, are not confounded; the auxiliary tones of the one are different from those of the other, and these latter tones, uniting themselves to the fundamental tones of the two instruments, destroy the identity of the sounds. It is a case of adding unequals to equals; the sums are unequal.

All bodies and instruments, then, employed for producing musical sounds emit, besides their fundamental tones, tones due to higher orders of vibration. The Germans embrace all such sounds under the general term *Obertöne.* I think it will be an advantage if we, in England, adopt the term *overtones* as the equivalent of the term employed in Germany. One has occasion to envy the power of the German language to adapt itself to requirements of this nature. The term *Klangfarbe,* for example, employed by Helmholtz is exceedingly expressive; and we need its equivalent also. You know that colour depends upon rapidity of vibration, that blue light bears to red the same relation that a high tone does to a low one. A simple colour has but one rate of vibration, and it may be regarded as the analogue of a simple tone in music. A *tone,* then, may be defined as the product of a vibration which cannot be decomposed into more simple ones. A compound colour, on the contrary, is produced by the admixture of two or more simple ones, and an assemblage of tones, such

* 'This quality of sound, sometimes called its register colour or timbre.' —THOMAS YOUNG, *Essay on Music.*

as we obtain when the fundamental tone and the harmonics of a string sound together, is called by the Germans a *Klang*. May we not employ the English word *clang* to denote the same thing, and thus give the term a precise scientific meaning akin to its popular one? And may we not, like Helmholtz, add the word *colour* or *tint*, to denote the character of the clang, using the term *clang-tint* as the equivalent of Klangfarbe?

With your permission I shall henceforth employ these terms; and now it becomes our duty to look a little more closely than we have hitherto done into the subdivision of a string into its harmonic segments. Our monochord with its stretched wire is before you. The scale of the instrument is divided into 100 equal parts. At the middle point of our wire stands the number 50; at a point almost exactly one-third of its length from its end stands the number 33; while at distances equal to one-fourth and one-fifth of its length from its end, stand the numbers 25 and 20 respectively. These numbers are sufficient for our present purpose. I pluck the wire at 50; you hear its clang, rather hollow and dull. I pluck it at 33; the clang is different. I pluck it at 25; the clang is different from either of the former. As I retreat from the centre of the string, the clang-tint becomes more brilliant, the sound more brisk and sharp. What is the reason of these differences in the sound of the same wire?

The celebrated Thomas Young, once professor in this Institution, enables us to solve the question. He proved that when any point of a string is plucked, all the higher tones which require that point *for a node* vanish from the clang. Let me illustrate this experimentally. I pluck the point 50, and permit the string to sound. And now I affirm that the first overtone, which corresponds to a division of the string into two vibrating parts, is absent from the clang. If it were present the damping of the point 50 would not interfere with it, for this point would be its node. I now damp the point 50; the fundamental tone is quenched, and no octave of that tone is heard. Along with the octave its whole progeny of overtones, with rates of vibration four times, six times, eight times —all even numbers of times—the rate of the fundamental tone, disappear from the clang. All these tones require a node at the centre, where, according to the principle of Young, it cannot now be formed. But I now pluck some other point, say 25, and damp 50 as before. The fundamental tone is gone, but its octave, clear and full, rings in your ears. The point 50 in this case, not being the one plucked, a node can form there ; it *has* formed, and the two halves of the string continue to vibrate after the vibrations of the string as a whole have been extinguished. Again I pluck the point 33, and am now sure that the second harmonic or overtone is absent from the clang. I prove this by damping the point 33. If the second harmonic were on the string this damping would not affect it, for 33 is the node of the harmonic. But no tone corresponding to a division of the string into three vibrating parts is now heard. The tone is not heard because it was never there.

All the overtones which depend on this division, those with six times, nine times, twelve times the rate of vibration of the fundamental one, are also withdrawn from the clang. I now pluck 20, and damp 33 as before; the damping in this case does not extinguish the second harmonic, which continues to sound clearly and fully after the extinction of the fundamental tone. In this case the point 33, not being that plucked, a node can form there, and the string can divide itself into three parts accordingly. In like manner, if I pluck 25 and damp 25 the third harmonic is not heard; but if I pluck a point between 25 and the end of the wire, and then damp 25, the third harmonic is plainly heard. And thus we might proceed, the general rule enunciated by Young, and illustrated by these experiments, being that when any point of a string is plucked or struck, or, as Helmholtz adds, agitated with a bow, the harmonic which requires that point for a node vanishes from the general clang of the string.

All this makes clear to you what a potent influence these higher vibrations must have upon the quality of the tone emitted by the string. The sounds which ring in your ears so plainly after the fundamental tone is quenched, mingled with that note before it was extinguished. It seems strange that tones of such power could be so masked by the fundamental one that even the disciplined ear of a musician is unable to separate the one from the other. But Helmholtz has made it perfectly clear that this is due to want of practice and attention. The musician's faculties were never exercised in this direction. There are numerous effects which the musician can distinguish, because his art demands the habit of distinguishing them. But it is no necessity of his art to resolve the clang of an instrument into its constituent tones. By attention, however, even the unaided ear can accomplish this, particularly if the mind be informed beforehand what the ear has to bend itself to find.

And this brings to my mind an occurrence which took place in this room at the beginning of my acquaintance with Mr. Faraday. I wished to show him a peculiar action of an electro-magnet upon a crystal. I had everything arranged, when just before I excited the magnet he laid his hand upon my arm and asked, 'What am I to look for?' Amid the assemblage of impressions connected with an experiment, even this prince of experimenters felt the advantage of having his attention directed to the special point in question. And such help is more particularly needful, with an effect so entangled and so intimately blended as the composite tones of a clang. One way of helping the attention when we desire to isolate a particular tone, is to sound that tone feebly on a string of its own. The ear having thus made the acquaintance of the tone, glides readily from it to the tone of the same pitch in a composite clang, and detaches it more readily from its companions. In the experiments executed a moment ago, where our aim in each respective case was to bring out the higher tone of the string in all its power, we entirely extinguished the fundamental tone. We can, however, enfeeble it without destroying it. I pluck this string at 33, and lay my feather lightly for a moment on the string at 50. I thereby lower the fundamental tone so much, that its octave can make itself plainly heard. Again, touching the string at 50, I lower the fundamental tone still more; so that now its first harmonic is more powerful than itself. You hear the sound of both, and you might have heard them in the first instance by a sufficient stretch of attention.

The harmonics of a string may be augmented or subdued within wide limits. They may, as we have seen, be masked by the fundamental tone, and they may also effectually mask it. A stroke with a hard body is favourable, while a stroke with a soft body is unfavourable to their development. They depend moreover on the promptness with which the body striking the string retreats after striking. Thus they are influenced by the weight and elasticity of the hammers in the pianoforte. They also depend upon the place at which the shock is imparted. When, for example, a string is struck in the centre, the harmonics are less powerful than when it is struck near one end. Pianoforte manufacturers have found that the most pleasing tone is excited when the point against which the hammer strikes is from $\frac{1}{7}$th to $\frac{1}{9}$th of the length of the wire from its extremity. Helmholtz, who is equally eminent as a mathematician and as an experimental philosopher, has calculated the theoretic intensity of the harmonics developed in various ways; that is to say, the actual *vis viva* or energy of the vibration, irrespective of its effects upon the ear. A single example given by him will suffice to illustrate this subject. Calling the intensity of the fundamental tone, in each case, 100, that of the second harmonic, when the string was simply pulled aside at a point $\frac{1}{7}$th of its length from its end, and then liberated, was found to be 56·1, or a little better than one-half. When the string was struck with the hammer of a pianoforte, whose contact with the string endured for $\frac{3}{7}$ths of the period of vibration of the fundamental tone, the intensity of the same tone was 9. In this case the second harmonic was nearly quenched. When, however, the duration of contact was diminished to $\frac{3}{20}$ths of the period of the fundamental, the intensity of the harmonic rose to 357; while, when the string was sharply struck with a very hard hammer, the intensity mounted to 505, or to more than quintuple that of the fundamental tone.*

But why should pianoforte makers strike the middle strings of their instruments at a point from $\frac{1}{7}$th to $\frac{1}{9}$th of their lengths from their extremities? They had no reason for doing so, beyond the fact that the striking of this point gave their ears the greatest satisfaction. The practice, however, is not without a reason. Up to the tones which require these points as nodes, as Helmholtz has pointed out, the overtones all form chords with the fundamental; but the sixth and eighth overtones of the wire do not enter into such chords; they are dissonant tones, and hence the desirability of doing away with them. This is accomplished by making the point at which a node is required

* *Lehre von den Tonempfindungen*, p. 135.

that on which the hammer falls. The possibility of the tone forming is thereby shut out, and its injurious effect avoided.

The sounds of the Eolian harp are produced by the division of suitably stretched strings into a greater or less number of harmonic parts by a current of air passing over them. The instrument is usually placed in a window between the sash and frame, so as to leave no way open to the entrance of the air except over the strings. Mr. Wheatstone recommends, as a good illustration of this point, the stretching of a first violin string at the bottom of a door which does not closely fit. When the door is shut, the current of air entering beneath sets the string in vibration, and when a fire is in the room, the vibrations are so intense that a great variety of sounds are simultaneously produced.* A gentleman in Basel once constructed with iron wires a large instrument which he called the weather-harp or giant-harp, and which, according to its maker, sounded as the weather changed. Its sounds were also said to be evoked by changes of terrestrial magnetism. Chladni pointed out the error of these notions, and reduced the action of the instrument to that of the wind upon its strings.

Finally, with regard to the vibrations of a wire, the experiments of Dr. Young, who was the first to employ optical methods in such experiments, must be mentioned. He allowed a sheet of sunlight to cross a pianoforte wire, and obtained thus a brilliant dot. Striking the wire he caused it to vibrate, the dot described a luminous line like that produced by the whirling of a burning coal in the air, and the form of this line revealed the character of the vibration. It was rendered manifest by these experi-

* The action of such a string is substantially the same as that of the syren. The string renders intermittent the current of air. Its action also resembles that of a *reed*. See Lecture V.

ments that the oscillations of the wire were not confined to a single plane, but that it described in its vibrations curves of greater or less complexity. Superposed upon the vibration of the whole string were partial vibrations, which revealed themselves as loops and sinuosities. Some of the lines observed by Dr. Young are given in

FIG. 48.

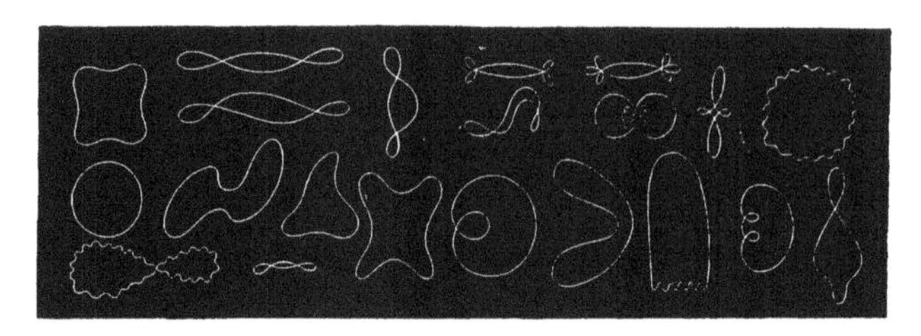

fig. 48. Every one of these figures corresponds to a distinct impression made by the wire upon the surrounding air. The form of the sonorous wave is affected by these superposed vibrations, and thus they influence the clang-tint or quality of the sound.

SUMMARY OF LECTURE III.

The amount of motion communicated by a vibrating string to the air is too small to be perceived as sound, even at a small distance from the string.

Hence, when strings are employed as sources of musical sounds, they must be associated with surfaces of larger area which take up their vibrations, and transfer them to the surrounding air.

Thus the tone of a harp, a piano, a guitar, or a violin, depends mainly upon the sound-board of the instrument.

The following four laws regulate the vibrations of strings:—The rate of vibration is inversely proportional to the length; it is inversely proportional to the diameter; it is directly proportional to the square root of the stretching weight or tension; and it is inversely proportional to the square root of the density of the string.

When strings of different diameters and densities are compared, the law is, that the rate of vibration is inversely proportional to the square root of the weight of the string.

When a stretched rope, or an india-rubber tube filled with sand, with one of its ends attached to a fixed object, receives a jerk at the other end, the protuberance raised upon the tube runs along it as a pulse to its fixed end, and, being there reflected, returns to the hand by which the jerk was imparted.

The time required for the pulse to travel from the hand to the fixed end of the tube and back, is that required by the whole tube to execute a complete vibration.

When a series of pulses are sent in succession along the tube the direct and reflected pulses meet, and, by their coalescence divide the tube into a series of vibrating parts, called *ventral segments*, which are separated from each other by points of apparent rest, called *nodes*.

The number of ventral segments is directly proportional to the rate of vibration at the free end of the tube.

The hand which produces these vibrations may move through less than an inch of space: while by the accumulation of its impulses, the amplitude of the ventral segments may amount to several inches, or even to several feet.

If an india-rubber tube, fixed at both ends, be encircled at its centre by the finger and thumb, when either of its halves is pulled aside and liberated, both halves are thrown into a state of vibration.

If the tube be encircled at a point one-third, one-fourth, or one-fifth of its length from one of its ends, on pulling the shorter segment aside and liberating it, the longer segment divides itself into two, three, or four vibrating parts, separated from each other by nodes.

The number of vibrating segments depends upon the rate of vibration at the point encircled by the finger and thumb.

Here also the amplitude of vibration at the place encircled by the finger and thumb may not be more than a fraction of an inch, while the amplitude of the ventral segments may amount to several inches.

A musical string damped by a feather at a point one-half, one-third, one-fourth, one-fifth, &c., of its length from one of its ends, and having its shorter segment agitated, divides itself exactly like the india-rubber tube. Its division may be rendered apparent by placing little paper riders across it. Those placed at the ventral segments are thrown off, while those placed at the nodes retain their places.

The notes corresponding to the division of a string into its aliquot parts are called the *harmonics* of the string.

When a string vibrates as a whole, it usually divides at the same time into its aliquot parts. Smaller vibrations are superposed upon the larger, the tones corresponding to those smaller vibrations, and which we have agreed to call overtones, mingling at the same time with the fundamental tone of the string.

The addition of these overtones to the fundamental tone determines the *timbre* or *quality* of the sound, or, as we have agreed to call it, the *clang-tint.*

It is the addition of such overtones to fundamental tones of the same pitch which enables us to distinguish the sound of a clarionet from that of a flute, and the sound of a violin from both. Could the pure fundamental tones of these instruments be detached, they would be undistinguishable from each other; but the different admixture of overtones in the different instruments renders their clang-tints diverse, and therefore distinguishable.

Instead of the heavy india-rubber tube in the experiments above referred to, we may employ light silk strings, and, instead of the vibrating hand, we may employ vibrating tuning-forks, and cause the strings to swing as a whole, or to divide themselves into any number of ventral segments. Effects of great beauty are thus obtained, and, by experiments of this character, all the laws of vibrating strings may be demonstrated.

When a stretched string is plucked aside, or agitated by a bow, all the overtones which require the point agitated for a node vanish from the clang of the string.

The point struck by the hammer of a piano is from one-seventh to one-ninth of the length of the string from its end: by striking this point, the notes which require it as a node cannot be produced, a source of dissonance being thus avoided.

LECTURE IV.

VIBRATIONS OF A ROD FIXED AT BOTH ENDS: ITS SUBDIVISIONS AND CORRESPONDING OVERTONES—VIBRATIONS OF A ROD FIXED AT ONE END—THE KALEIDOPHONE—THE IRON FIDDLE AND MUSICAL BOX—VIBRATIONS OF A ROD FREE AT BOTH ENDS—THE CLAQUEBOIS AND GLASS HARMONICA—VIBRATIONS OF A TUNING-FORK: ITS SUBDIVISION AND OVERTONES—VIBRATIONS OF SQUARE PLATES—CHLADNI'S DISCOVERIES—WHEATSTONE'S ANALYSIS OF THE VIBRATIONS OF PLATES—CHLADNI'S FIGURES—VIBRATIONS OF DISCS AND BELLS—EXPERIMENTS OF FARADAY AND STREHLKE.

OUR last lecture was devoted to the transverse vibrations of strings. I propose devoting the present lecture to the transverse vibrations of rods, plates, and bells, commencing with the case of a rod fixed at both ends. Its modes of vibration are exactly those of a string. It vibrates as a whole, and can also divide itself into two, three, four, or more vibrating parts. But, for a reason to be immediately assigned, the laws which regulate the pitch of the successive notes are entirely different in the two cases. Thus, when a string divides into two equal parts, each of its halves vibrates with twice the rapidity of the whole;

Fig. 49.

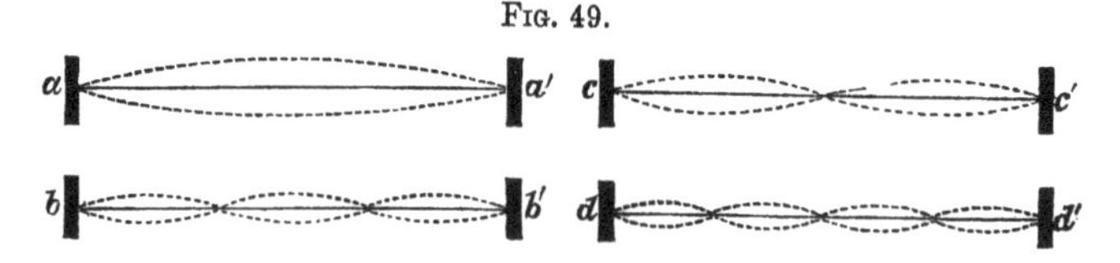

while, in the case of the rod, each of its halves vibrates with nearly three times the rapidity of the whole. With greater strictness, the ratio of the two rates of vibration is as 9 : 25 that is to say, as the square of 3 to the square of 5. In fig. 49, *aa'*, *cc'*, *bb'*, *dd'*, I have sketched the first four modes of vibration of a rod fixed at both ends: the successive rates of vibration in the four cases bear to each other the following relation :—

Number of nodes . .	0	1	2	3
Number of vibrations .	9	25	49	81

the last row of figures being the squares of the odd numbers 3, 5, 7, 9.

In the case of a string, the vibrations are maintained by a tension externally applied; in the case of a rod, the vibrations are maintained by the elasticity of the rod itself. The modes of division are in both cases the same, but the forces brought into play are different, and hence also the successive rates of vibration.

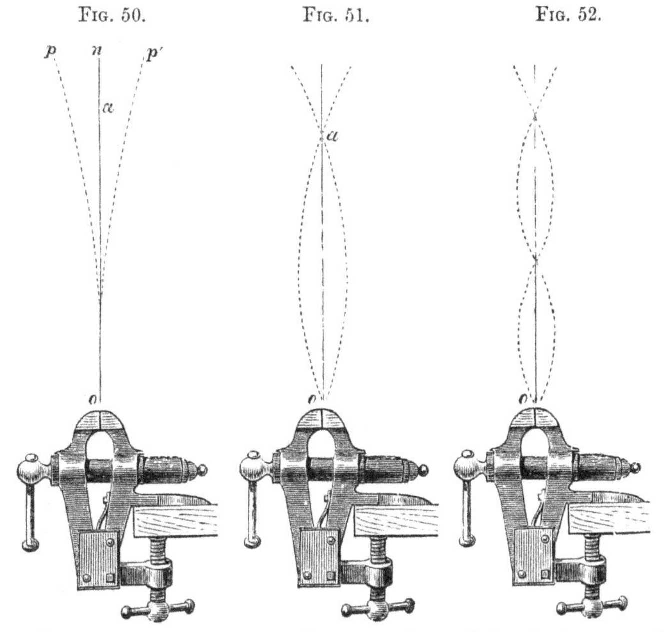

FIG. 50. FIG. 51. FIG. 52.

Let us now pass on to the case of a rod fixed at one end and free at the other. Here also it is the elasticity of the material, and not any external tension, that sustains the vibrations. Approaching, as usual, the sonorous through the more grossly mechanical, I fix this long rod of iron, *n o*, fig. 50, in a vice, draw it aside and liberate it. To make its vibrations more evident, I throw, by the electric light, its shadow upon a screen. It oscillates as a whole to and fro, between the points *p p′*. But the rod is capable of other modes of vibration. I damp it at the point *a*, by holding it gently there between my finger and thumb, and strike it sharply between *a* and *o*. The rod divides into two vibrating parts, separated by a node fig. 51. You see upon the screen this shadowy spindle between *a* and the vice below, and this shadowy fan above *a*, with this black node between both. The division may be effected without damping *a*, by merely imparting a sufficiently sharp shock to the rod between *a* and *o*. In this case, however, besides oscillating in parts, the rod oscillates as a whole, the partial oscillations being superposed upon the large one. You notice, moreover, that the amplitude of the partial oscillations depends upon the promptness of my stroke. When the stroke is sluggish the partial division is but feebly pronounced, the whole oscillation being most marked. But when the shock is sharp and prompt, the whole oscillation is feeble, and the partial oscillations are executed with vigour. Now if the vibrations of this rod were rapid enough to produce a musical sound, the oscillation of the rod as a whole would correspond to its fundamental tone, while the division of the rod into two vibrating parts would correspond to the first of its overtones. If, moreover, the rod vibrated as a whole and as a divided rod at the same time, the fundamental tone and the overtone would be heard simultaneously. By damping the proper point and imparting the proper shock, I can still further subdivide the rod, as shown in fig. 52.

And now let us shorten our rod so as to bring its vibrations into proper relation to our auditory nerves.

It is now about four inches long; across its upper end I draw a bow, and you hear a low musical sound. I shorten the rod still more; the tone is higher; and by continuing to shorten I augment the speed of vibration, until finally the sound becomes painfully acute. These musical vibrations differ only in rapidity from the grosser oscillations which a moment ago appealed to the eye.

The increase in the rate of vibration here observed is ruled by a definite law; the number of vibrations executed in a given time is inversely proportional to the square of the length of the vibrating rod. You hear the sound of this strip of brass, two inches long, as I pass the fiddle bow over its end. I make the length of the strip one inch; the sound is now the double octave of the last one; its rate of vibration is four times as rapid. Thus by doubling the length of the vibrating strip we reduce its rate of vibration to one-fourth; by trebling the length we reduce the rate of vibration to one-ninth; by quadrupling the length we reduce the vibrations to one-sixteenth, and so on. It is plain that by proceeding in this way we should finally reach a length where the vibrations would be sufficiently slow to be counted by the eye. Or, that beginning with a long strip, whose vibrations could be counted, we might, by shortening, not only make the strip sound, but also determine the rates of vibration corresponding to its different tones. Supposing we start with a strip 36 inches long, which vibrates once in a second, the strip reduced to 12 inches would, according to the above law, execute 9 vibrations a second: reduced to 6 inches it would execute 36, to 3 inches, 144; while if reduced to 1 inch in length it would execute 1,296 vibrations in a second. It is easy to fill the spaces between the lengths here given, and thus to determine the rate of vibration corresponding to any particular tone. This method was proposed and carried out by Chladni.

A musical instrument may be formed of short rods. Into this common wooden tray a number of pieces of stout iron wire of different lengths are fixed, being ranged in a semicircle. I pass the fiddle bow over the series, and obtain a succession of very pleasing notes. A competent performer could certainly extract very tolerable music from a sufficient number of these iron pins. The iron fiddle (*violon de fer*) is thus formed. The notes of the ordinary musical box are also produced by the vibrations of tongues of metal fixed at one end. Pins are fixed in a revolving cylinder, the free ends of the tongues being lifted by these pins and then suddenly let go. The tongues vibrate, their length and strength being so arranged as to produce in each particular case the proper rapidity of vibration.

Mr. Wheatstone has devised a simple and ingenious optical method for the study of vibrating rods fixed at one end. Attaching light glass beads, silvered within, to the end of a metal rod, and allowing the light of a lamp or candle to fall upon the bead, he obtained a small intensely illuminated spot. When the rod vibrated, this spot described a brilliant line which showed the character of the vibration. A knitting needle fixed in a vice with a small bead stuck on to it by marine glue answers perfectly as an illustration. In Mr. Wheatstone's more complete instrument, which he calls a *kaleidophone*, the vibrating rods are firmly screwed into a massive stand. Extremely beautiful figures are obtained by this simple contrivance, some of which I will endeavour to project on a magnified scale upon the screen before you.

Fixing the rod horizontally in the vice, I permit the condensed beam of the electric lamp to fall upon the silvered bead, and thus obtain a spot of sunlike brilliancy. Placing a lens in front of the bead, I throw a bright image of the spot upon the screen, and drawing the needle aside, suddenly liberate it. The spot describes a ribbon of light at first straight, but speedily opening out into an ellipse, passing into a circle, and then again through a second ellipse back to a straight line. This is due to the fact that a rod held thus in a vice vibrates not only in the direction in which it is drawn aside, but also at right angles to this direction. The curve is due to the combination of two rectangular vibrations.* I now wish to show you, that while the rod is thus swinging as a whole it may also divide into vibrating parts. By properly drawing a violin bow across the needle I obtain this serrated circle, fig. 53, a number of small undulations being superposed upon the larger one. You moreover hear a musical tone, which you did not hear when the rod vibrated as a whole only; its oscillations, in fact, were then too slow to excite such a tone. The vibrations which produce these sinuosities, and which correspond to the first division of the rod, are executed with about $6\frac{1}{4}$ times the rapidity of the vibrations of the rod swinging as a whole. Again I draw the bow; the note rises in pitch, the serrations run more closely together, forming on the screen a luminous ripple more minute and, if possible, more exquisitely beautiful than the last one, fig. 54. Here we have the second division of the rod, the sinuosities of which correspond to $17\frac{13}{36}$ times its rate of vibration as a whole. Thus every change in the sound of the rod is accompanied by a change of the figure upon the screen.

FIG. 53.

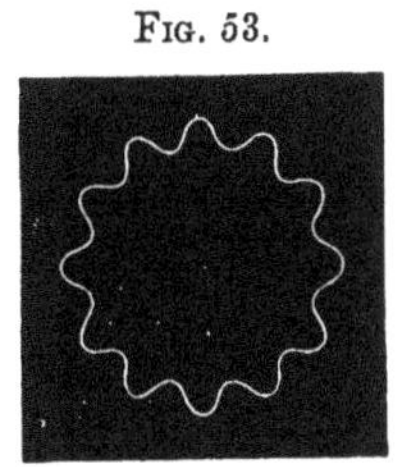

FIG. 54.

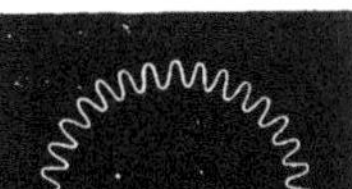

The rate of vibration of the rod as a whole, is to the rate corresponding to its first division, nearly as the square of 2 is

* Chladni also observed this compounding of vibrations, and executed a series of experiments, which, in their developed form, are those of the kaleidophone. The composition of vibrations will be studied at some length in a subsequent lecture.

to the square of 5, or as 4 : 25. From the first division onwards the rates of vibration are approximately proportional to the squares of the series of odd numbers 3, 5, 7, 9, 11, &c. Supposing the vibrations of the rod as a whole to number 36, then the vibrations corresponding to this and to its successive divisions would be expressed approximately by the following series of numbers : —

36, 225, 625, 1225, 2025, &c.

In fig. 55, *a*, *b*, *c*, *d*, *e*, are shown the modes of division corresponding to this series of numbers. You will not fail to observe that these overtones of a vibrating rod rise far more rapidly in pitch than the harmonics of a string.

Fig. 55.

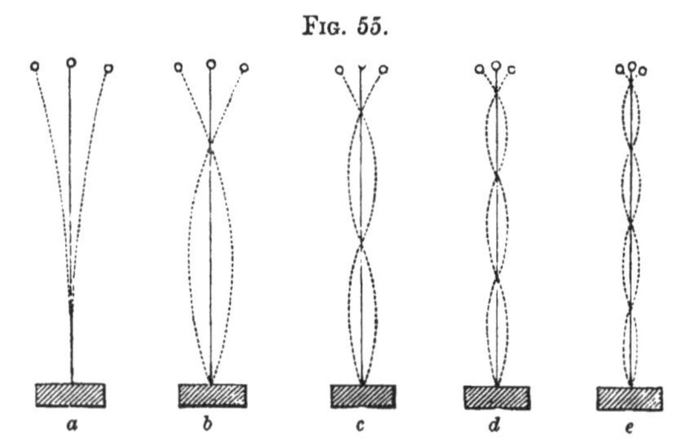

Other forms of vibration may be obtained by smartly striking the rod with the finger near its fixed end. In fact an almost infinite variety of luminous scrolls can be thus produced, the beauty of which may be inferred from the subjoined figures first obtained by Mr. Wheatstone. They may be produced by illuminating the bead with sunlight, or with the light of a lamp or candle. The scrolls, moreover, may be doubled by employing two candles instead of one. Two spots of light then appear, each of which describes its own luminous line when the knitting needle is set in vibration. In a subsequent lecture we shall become acquainted with Mr. Wheatstone's application of his method to the study of rectangular vibrations.

FIG. 56.

From a rod or bar fixed at one end, we will now pass to rods or bars free at both ends; for such an arrangement has also been employed in music. By a method afterwards to be described, Chladni, the father of modern acoustics, determined experimentally the modes of vibration possible to such bars. The simplest mode of division here possible occurs when the rod is divided by two nodes into three vibrating parts. This mode is easily illustrated by this flexible box ruler, six feet long. Holding it at about twelve inches from its two ends between the forefinger and thumb of each hand, and shaking it, or causing its centre to be struck, it vibrates, the middle segment forming a shadowy spindle, and the two ends forming fans. The shadow of the ruler on the screen renders the mode of vibration still more evident. In this case the distance of each node from the end of the ruler is about one-fourth of the distance between the two nodes. In its second mode of vibration the rod or ruler is divided into four vibrating parts by three nodes. In fig. 57, 1 and 2, these respective modes

Fig. 57.

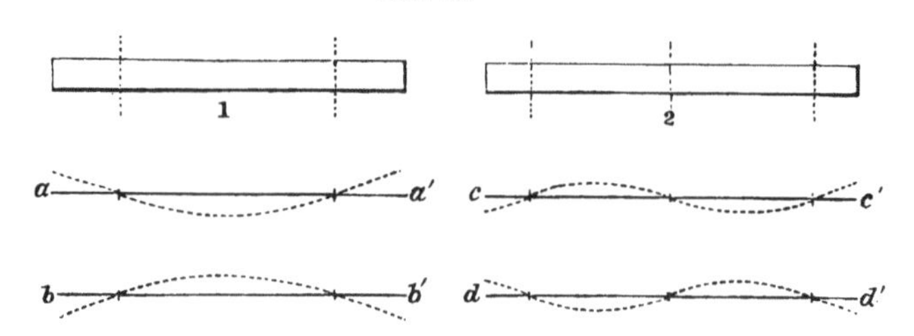

of division are shown. Looking at the edge of the bar 1, the dotted lines *a a′*, *b b′* show the manner in which the segments bend up and down when the first division occurs, while *c c′*, *d d′* show the mode of vibration corresponding to the second division. The deepest tone of a rod free at both ends is higher than the deepest tone of a rod fixed at one end in the proportion of 4 : 25. Beginning with the first two nodes, the rates of vibration of the free bar rise in the following proportion:—

Number of nodes	2, 3, 4, 5, 6, 7.
Numbers to the squares of which the pitch is approximately proportional	3, 5, 7, 9, 11, 13.

Here also we have a similarly rapid rise of pitch to that noticed in the last two cases.

For musical purposes the first division only of a free

rod has been employed. When bars of wood of different lengths, widths, and depths, are strung along a cord which passes through the nodes we have the *claque-bois* of the French, an instrument now before you, A B, fig. 58. Supporting the cord at one end by a hook *k*, and holding it at the other in my left hand, I run the hammer *h* along the series of bars, and produce this agreeable succession of musical tones. Instead of using the cord, the bars may rest at their nodes on cylinders of twisted straw; hence the name straw-fiddle sometimes applied to this instrument. Chladni

Fig. 58.

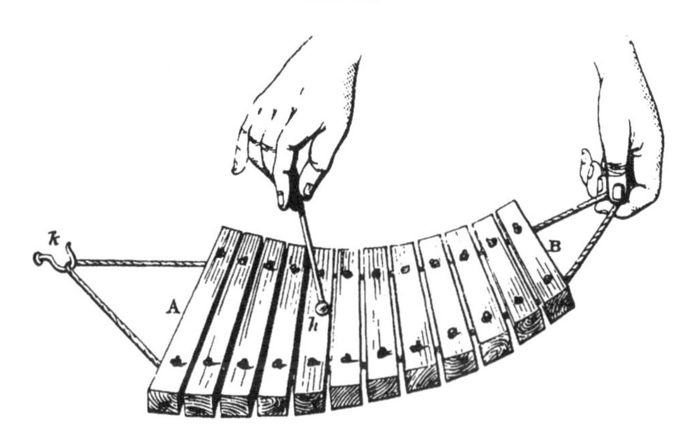

informs us that it is introduced as a play of bells (Glockenspiel) into Mozart's opera of the Zauberflöte. If, instead of bars of wood, we employ strips of glass, we have the glass harmonica.

From the vibrations of a bar free at both ends, it is easy to pass to the vibrations of a tuning-fork, as analysed by Chladni. Supposing *a a*, fig. 59, to represent a straight steel bar, with the nodal points corresponding to its first mode of division marked by the transverse dots. Let the bar be bent to the form *b b*; the two nodal points still remain, but they have approached nearer to each other. The tone of the bent bar is also somewhat lower than that of the straight one. Passing through various stages of bending, *c c*, *d d*, we at length convert the bar into a tuning-fork *e e*, with parallel prongs; it still retains its two nodal points, which, however, are much closer together than when the bar was straight. When such a fork sounds its deepest note, its free ends oscillate, as in fig. 60, where the prongs vibrate between the limits *b n*, and *f m* and where *p* and *q* are the nodes. There is no division of a tuning-fork corresponding to the division of a straight bar by three nodes. first overtone of the fork, we have a node on each prong, and In its second mode of division, which corresponds to the

Fig. 59. Fig. 60.

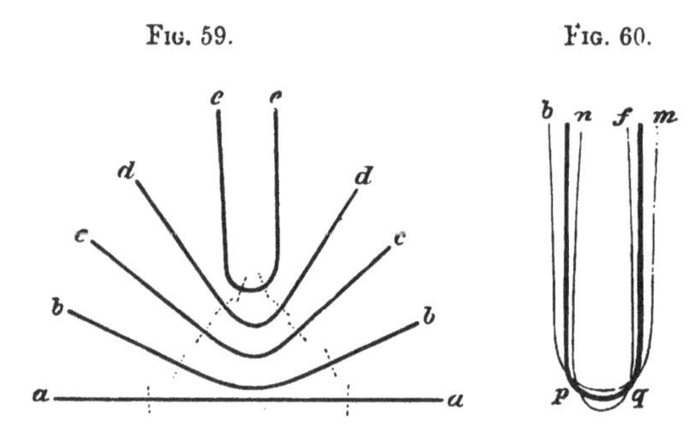

two at the bottom. The principle of Young referred to at page 118 extends to tuning-forks. If you wish to have the fundamental tone free from an overtone, you draw your bow across the fork at the place where a node is required by the latter. In the third mode of division there are two nodes on each prong, and one at the bottom. In the fourth division there are two nodes on each prong, and two at the bottom; while in its fifth division there are three nodes on each prong and one at the bottom. The first overtone of the fork requires, according to Chladni, $6\frac{1}{4}$ times the number of vibrations of the fundamental tone. It is easy to elicit the overtones of tuning-forks. Here, for example, is our old series, vibrating respectively 556, 320, 384, and 512 times in a second. I pass from the fundamental tone to the first overtone of each; you notice that the interval is vastly greater than that between the fundamental tone and the first overtone of a stretched string. From the numbers just mentioned we pass at once to 1600, 2000, 2400, and 3200 vibrations a second. Chladni's numbers, however, though approximately correct, are not always rigidly verified by experiment. A pair of forks, for example, may have their fundamental tones in perfect unison, and their overtones discordant. Two such forks are now before you. I sound the fundamental tones of both; the unison is perfect. I now sound the first overtones of both; they are not in unison: you hear rapid 'beats,' which grate upon the ear. By loading one of the forks with wax, I can bring the two overtones into unison; but now the fundamental tones produce loud beats when sounded together. This could not occur if the first overtone of each fork was produced by a number of vibrations exactly $6\frac{1}{4}$ times the rate of its fundamental. In a series of forks examined by Helmholtz, the number of vibrations of the first overtone was from 5·8 to 6·6 times that of the fundamental. Starting from this first overtone, and including it, the rates of vibration of the whole series of overtones are as the squares of the numbers 3, 5, 7, 9, &c. That is to say, in the time required by the first overtone to execute 9 vibrations, the second executes 25, the third 49, the fourth 81, and so on. Thus the overtones of the fork rise with far greater rapidity than those of a string. They also vanish more speedily, and hence adulterate to a less extent the fundamental tone by their admixture.

The device of Chladni for rendering these sonorous vibrations visible has been of immense importance to the science of acoustics. Lichtenberg had made the experiment of scattering an electrified powder over an electrified resin cake; the arrangement of the powder revealing the electric condition of the surface. This experiment suggested to Chladni the idea of rendering sonorous vibrations visible by means of sand strewn upon the surface of the vibrating body. Chladni's own account of his discovery is of sufficient interest to justify its introduction here.

'As an admirer of music, the elements of which I had begun to learn rather late, that is, in my nineteenth year, I noticed that the science of acoustics was more neglected than most other portions of physics. This excited in me the desire to make good the defect, and by new discovery to render some service to this part of science. In 1785 I had observed that a plate of glass or metal gave different sounds when it was struck at different places, but I could nowhere find any information regarding the corresponding modes of vibration. At this time there appeared in the journals some notices of an instrument made in Italy by the Abbé Mazzocchi, consisting of bells, to which one or two violin bows were applied. This suggested to me the idea of employing a violin bow to examine the vibrations of different sonorous bodies. When I applied the bow to a round plate of glass fixed at its middle it gave different sounds, which, compared with each other, were (as regards the number of their vibrations) equal to the squares of 2, 3, 4, 5, &c.; but the nature of the motions to which these sounds corresponded, and the means of producing each of them at will, were yet unknown to me. The experiments on the electric figures formed on a plate of resin, discovered and published by Lichtenberg, in the memoirs of the Royal Society of Göttingen, made me presume that the different vibratory motions of a sonorous plate might also present different appearances, if a little sand or some other similar substance were spread on the surface. On employing this means, the first figure that presented itself to my eyes upon the circular plate already mentioned, resembled a star with ten or twelve rays, and the very acute sound, in the series alluded to, was that which agreed with the square of the number of diametral lines.'

I will now illustrate the experiments of Chladni, commencing with a square plate of glass held by a suitable clamp at its centre. I might hold the plate with my finger and thumb, if they could only reach far enough. Scattering fine sand over the plate, I damp the middle

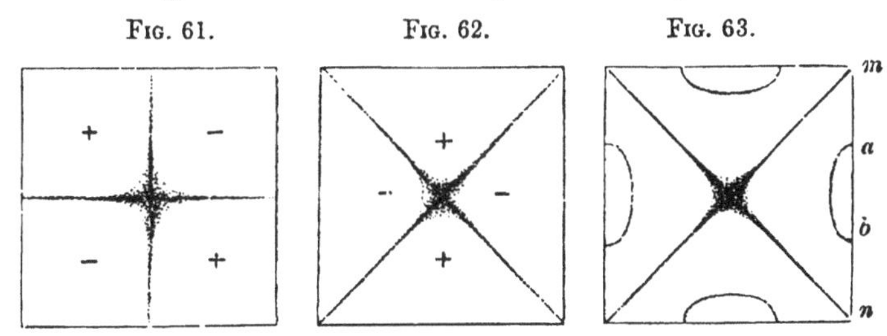

point of one of its edges by touching it with my finger nail, and draw a bow across the edge of the plate, near one of its corners. The sand is tossed away from certain parts of the surface, and collects along two *nodal lines* which divide the large square into four smaller ones, fig. 61. This division of the plate corresponds to its deepest tone.

The signs + and − here employed denote that the two squares on which they occur are always moving in opposite directions. When the squares marked + are above the average level of the plate those marked − are below it; and when those marked − are above the average level those marked + are below it. The nodal lines mark the boundaries of these opposing motions. They are the places of transition from the one motion to the other, and are therefore unaffected by either.

Scattering sand once more over its surface, I damp one of the corners of the plate, and excite it by drawing the bow across the middle of one of its sides. The sand dances over the surface of the plate, and finally ranges itself in two sharply-defined ridges along its diagonals, fig. 62. The note here produced is a fifth above the last. Again damping the points *a* and *b*, fig. 63, and drawing the bow across the centre of the opposite side of the plate, we obtain a far shriller note than in either of the former cases, and the manner in which the plate vibrates to produce this note is represented in the figure.

Fig. 64.

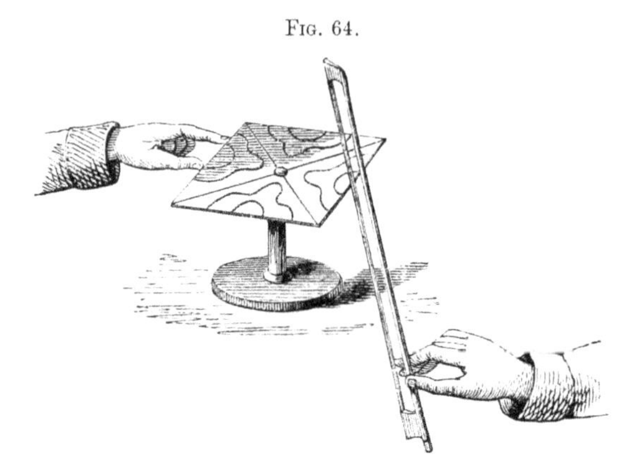

I have thus far employed plates of glass held by a clamp at the centre. Plates of metal are also suitable for such experiments. Fig. 64 represents a plate of brass, 12 inches square, and supported on a suitable stand. Damping it with the finger and thumb of my left hand at two points of its edge, and drawing the bow with my right across a vibrating portion of the opposite edge, I obtain this complicated figure.

Here follows a series of the beautiful patterns obtained by Chladni, by damping and exciting square plates in different ways. It is not only interesting but startling to see the suddenness with which these sharply defined figures are formed by the sweep of the bow of a skilful experimenter.

FIG. 65.

And now let us look a little more closely into the mechanism of these vibràtions. You know the manner in which a bar free at both ends divides itself when it vibrates transversely. A rectangular piece of glass or of sheet metal—the glass strips of the harmonica, for example, also obey the laws of rods and bars, vibrating with their ends free. In fig. 66 is drawn a rectangle *a*, with the nodes corresponding to its first division marked upon it, and underneath it is placed a figure showing the manner in which the rectangle, looked at edgeways, bends up and down when it is set in vibration.* For the sake of plainness the bending is greatly exaggerated. The figures *b* and *c* indicate that the vibrating parts of the plate alternately rise above and fall below the average level of the plate. At one moment, for example, the centre of the plate is above the level, and its ends below it, as at *b*; while at the next moment its centre is below, and its two ends above the average level, as at *c*. The vibrations of the plate consist in the quick and successive assumption of these two positions. Similar remarks apply to all other modes of division.

FIG. 66.

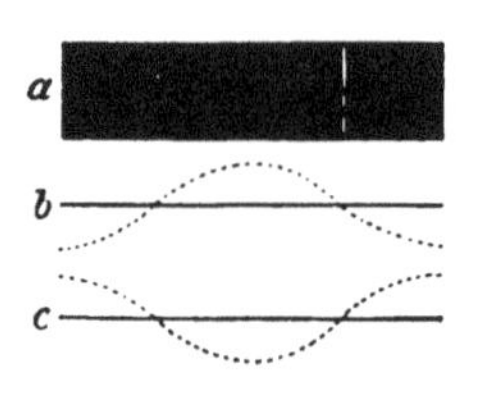

Now suppose the rectangle gradually to widen, till it becomes a square. There would then be no reason why the nodal lines should form parallel to one pair of sides rather than to the other. Let us now examine what would be the effect of the coalescence of two such systems of vibrations.

* I copy this figure from Mr. Wheatstone's memoir; the nodes, however, ought to be nearer the ends, and the free terminal portions of the dotted lines ought not to be bent upwards or downwards. The nodal lines in the next two figures are also drawn too far from the edges of the plates.

To keep your conceptions clear, take two squares of glass and draw upon each of them the nodal lines belonging to a rectangle. Draw the lines on one plate in white and on the other in black; this will help you to keep the plates distinct in your mind as you look at them. Now lay one square upon the other so that their nodal lines shall coincide, and then realise with perfect mental clearness both plates in a state of vibration. Let us assume, in the first instance, that the vibrations of the two plates are concurrent; that the middle segment and the end segments of each rise and fall together; and now suppose the vibrations of one plate transferred to the other. What would be the result? Evidently vibrations of a double amplitude on the part of the plate which has received this accession. But suppose the vibrations of the two plates, instead of being concurrent, to be in exact opposition to each other—that when the middle segment of the one rises the middle segment of the other falls—what would be the consequence of adding them together? Evidently a neutralisation of all vibration.

Instead of placing the plates so that their nodal lines coincide, set these lines now at right angles to each other. That is to say, push A over A′, fig. 67. In these figures the

FIG. 67.

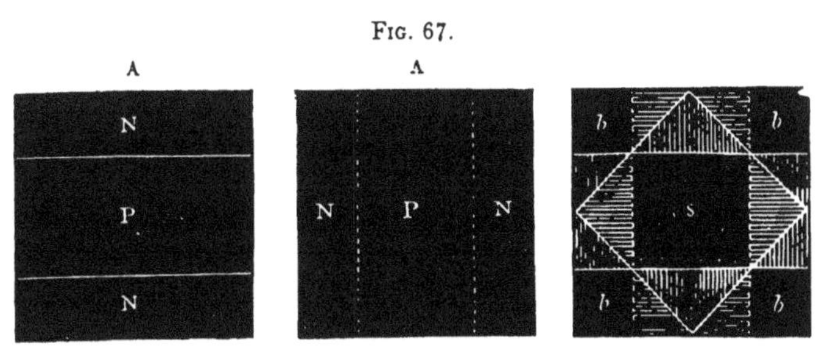

letter P means positive, indicating, in the section where it occurs, a motion of the plate upwards; while N means negative, indicating, where it occurs, a motion downwards. You have now before you a kind of check pattern, as shown in the third square, consisting of a square *s* in the middle, a smaller square *b* at each corner, and four rectangles at the middle portions of the four sides. Let the plates vibrate, and let the vibrations of their corresponding sections be concurrent, as indicated by the letters P and N; and then suppose the vibrations of one of them transferred to the other. What must result? A moment's reflection will show you that the big middle square will vibrate with twice its former energy; the same is true of the four smaller squares at the four corners; but you will at once convince yourselves that the vibrations in the four rectangles are in opposition, and that where their amplitudes are equal, they will destroy each other. The middle point of each side of the plate of glass is, therefore, a point of rest; the points where the nodal lines of the two plates cross each other are also points of rest. Draw a line through every three of these points and you obtain a second square inscribed in the first. The sides of this square are lines of no motion.

We have thus far been theorising; but I now clip this square plate of glass at a point near the centre of one of its edges, and drawing the bow across the adjacent corner of the plate, I obtain, when the glass is homogeneous, a close approximation to this inscribed square. The reason is that when the plate is agitated in this manner, the two sets of vibrations which we have been considering actually co-exist in the plate, and produce the figure due to their combination.

Again, place the squares of glass one upon the other exactly as in the last case; but now, instead of supposing them to concur in their vibrations, let their corresponding sections oppose each other: that is, let A cover A′, fig. 68. Then it is manifest that on superposing the vibrations the middle point of our middle square must be a point of rest; for here the vibrations are equal and opposite. The intersections of the nodal lines are also points of rest, and so also is every corner of the plate itself, for here the added vibrations are also equal and opposite. We have thus fixed four points of rest on each

FIG. 68.

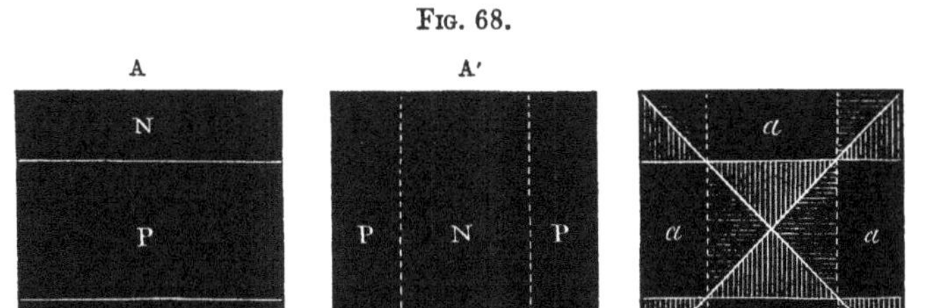

diagonal of the square. Draw the diagonals and they will represent the nodal lines consequent on the superposition of the two vibrations.

These two systems actually co-exist in the same plate when the centre is clamped and one of the corners touched, while the fiddle-bow is drawn across the middle of one of the sides. In this case the sand which marks the lines of rest arranges itself along the diagonals. I have thus endeavoured to put before you in the simplest possible manner a specimen of Mr. Wheatstone's analysis of these superposed vibrations.

Passing from square plates to round ones, we also obtain various beautiful effects. Here, for instance, is a disc of brass supported horizontally upon this upright stand. The disc is black, and I scatter fine white sand lightly over it. It is capable of dividing itself in various ways, and of emitting notes of various pitch. I will first of all sound the lowest fundamental note of the disc; this I do by touching its edge at a certain point, and by drawing the bow across the edge at a point 45 degrees distant from the damped one. You hear the note and you see the sand. It quits the four quadrants of the disc, and ranges itself along two of the diameters, fig. 69 A. When a disc divides itself thus into four vibrating segments, it sounds its deepest note. I now stop the vibration, clear the disc, and once more scatter sand over it. Damping its edge, and drawing the bow across it at a point 30 degrees distant from the damped one, the sand

Fig. 69.

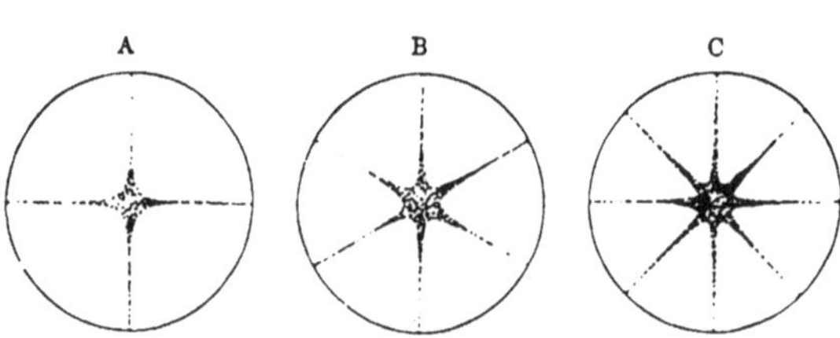

immediately arranges itself in a star. We have here six vibrating segments, separated from each other by their appropriate nodal lines, fig. 69 B. Again I damp a point, and agitate another nearer to the damped one than in the last instance; the disc divides itself into eight vibrating segments with lines of sand between them, fig. 69 C. In this way I continue to subdivide the disc into ten, twelve, fourteen, sixteen sectors, the number of sectors being always an *even* one. As the division becomes more minute, the vibrations become more rapid, and the pitch consequently more high. The note emitted by the sixteen segments into which the disc is now divided is so acute as to be almost painful to the ear. Here you have Chladni's first discovery. You can understand his emotion on witnessing this wonderful effect, 'which no mortal had previously seen. By rendering the centre of the disc free, and damping appropriate points of the surface, nodal circles and other curved lines may be obtained.

The rate of vibration of a disc is directly proportional to its thickness, and inversely proportional to the square of its diameter. I have here three discs; two of them of the same diameter, but one twice as thick as the other, and two of them of the same thickness, but one with twice the diameter of the other. According to the law just enunciated, the rates of vibration of these three discs are as the numbers 1, 2, 4. I sound them in succession, and the musical ears present can testify that they really stand to each other in the relation of a note to its octave and its double octave.

The actual movement of the sand towards the nodal lines may be studied by clogging the sand with a semifluid substance. Here are some specimens, in which gum has been employed to retard the motion of the particles. The curves which they individually describe are very clearly drawn upon the plates. M. Streblke has sketched these appearances, and from him I borrow the figures, A, B, C, fig. 70.

Fig. 70.

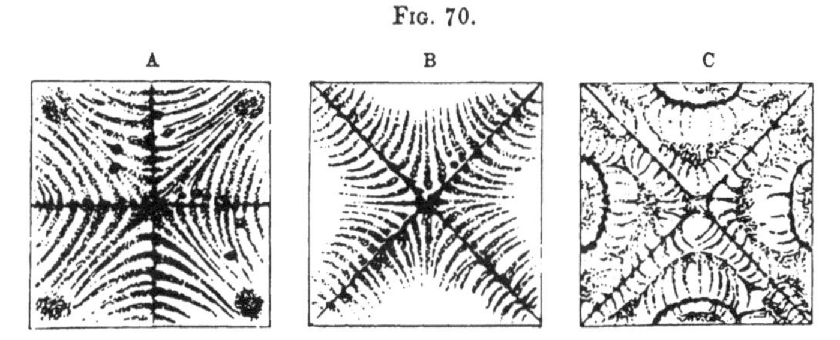

An effect of vibrating plates which long perplexed experimenters is here to be noticed. Mingled with the sand strewn over this plate is a little fine dust: in fact, to show you the effect more clearly, I have intentionally mixed with the sand the fine seed of lycopodium. This light substance, instead of collecting along the nodal lines, forms little heaps at the places of most violent motion. You see these heaps at the four corners of the plate, fig. 71, at the four sides of the plate, fig. 72, and lodged between the nodal lines of the plate, fig. 73. These three figures represent the three states of vibration illustrated in figs. 61, 62, and 63. The dust chooses in all cases the place of greatest agitation. Various explanations of this effect had been given, but it was reserved for Mr. Faraday to assign its extremely simple cause. The light powder is entangled by the little whirlwinds of air produced by the vibrations of the plate: it cannot escape from the little cyclones, though the heavier sand particles are

Fig. 71. Fig. 72. Fig. 73.

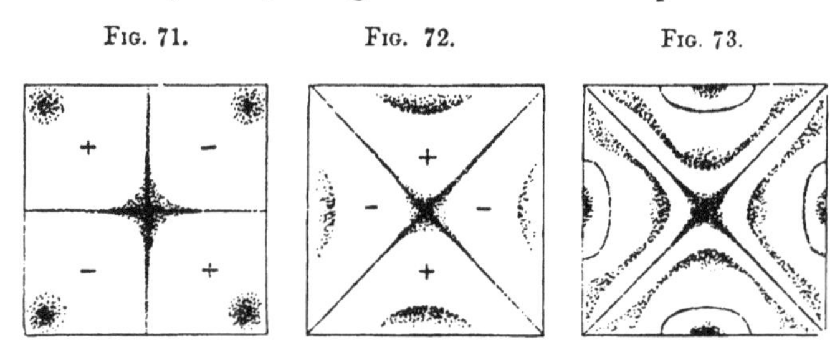

readily driven through them. When therefore the motion ceases, the light powder settles down in heaps at the places where the vibration was a maximum. In vacuo no such effect is observed: here all powders, light and heavy, move to the nodal lines.

The vibrating segments and nodes of a bell are similar to those of a disc. When a bell sounds its deepest note, the coalescence of its pulses causes it to divide into four vibrating segments, separated from each other by four nodal lines, which run up from the sound-bow to the crown of the bell. The place where the hammer strikes is always the middle of a vibrating segment; the point diametrically opposite is also the middle of such a segment. Ninety degrees from these points we have also vibrating segments, while at 45 degrees right and left of them we come upon the nodal lines. Supposing the strong dark circle in fig. 74 to represent the circumference of the bell in a state of quiescence, then when the hammer falls on any one of the segments a, c, b, or d, the sound-bow passes periodically through the changes indicated by the dotted lines. At one moment it is an oval, with $a\ b$

for its longest diameter; at the next moment it is an oval, with *c d* for its longest diameter. The changes from one oval to the other, constitute in fact the vibrations of the bell. The four points *n*, *n*, *n*, *n*, where the two ovals intersect each other, are the nodes. As in the case of a disc, the number of vibrations executed by a bell in a given time, varies directly as the square of the thickness, and inversely as the bell's diameter.

FIG. 74.

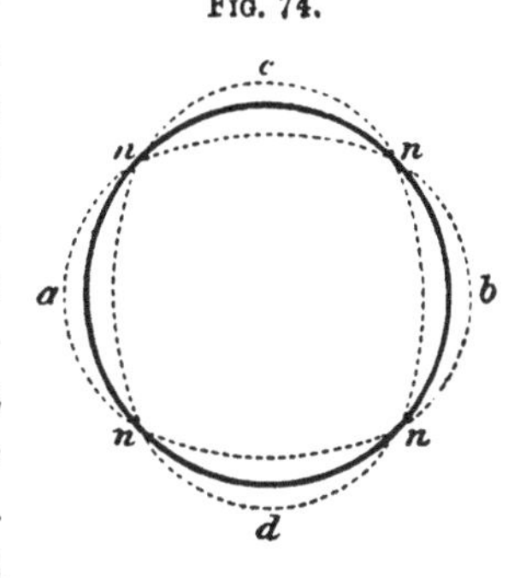

Like a disc, also, a bell can divide itself into any even number of vibrating segments, but not into an odd number. By damping proper points in succession, the bell can be caused to divide into 6, 8, 10, and 12 vibrating parts. Beginning with the fundamental note, the number of vibrations corresponding to the respective divisions of a bell, as of a disc, is as follows:—

Number of divisions . . .	4, 6, 8, 10, 12.
Numbers the squares of which express the rates of vibration	2, 3, 4, 5, 6.

Thus if the vibrations of the fundamental tone be 40, that of the next higher tone will be 90, the next 160, the next 250, the next 360, and so on. If the bell be thin the tendency to subdivision itself is so great, that it is almost impossible to bring out the pure fundamental tone without the admixture of the higher ones.

I will now repeat before you a homely, but an instructive experiment. This common jug, when I draw a fiddle-bow across its edge, divides into four vibrating segments, exactly like a bell. The jug is provided with a handle; and I wish you to notice the influence of this handle upon the tone. I draw the fiddle-bow across the edge at a point diametrically opposite to the handle. I do the same at a point 90° from the handle; the pitch of the note in both these cases is the same. In both cases the handle occupied the middle of a vibrating segment, loading that segment by its weight. But I now draw the bow at an angular distance of 45° from the handle; the note is sensibly higher than before. The handle in this experiment occupies a node; it no longer loads a vibrating segment, and hence the elastic force, having to cope with less weight, produces a more rapid vibration. The experiment here made with a jug Chladni executed with a tea-cup. Now bells often exhibit round their sound-bows an absence of

FIG. 75.

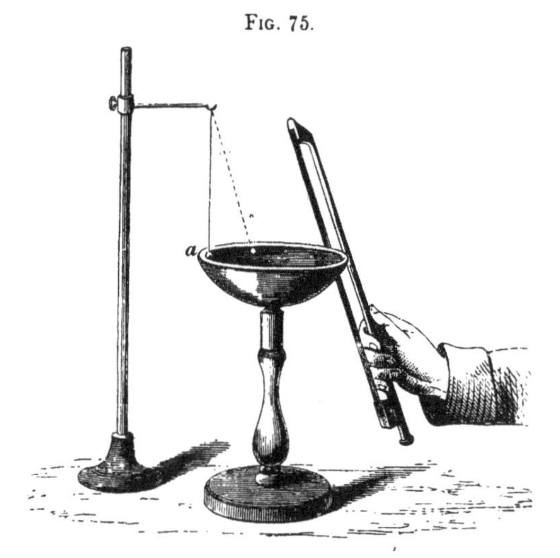

uniform thickness, tantamount to the want of symmetry in the case of our jug; and we shall learn subsequently, that the intermittent sound of many bells, noticed more particularly when their tones are dying out, is produced by the combination of two distinct rates of vibration, which have this absence of uniformity for their origin.

There are no points of absolute rest in a vibrating bell,

for the nodes of the higher tones are not those of the fundamental one. But that the various parts of the sound-bow, when the fundamental tone is predominant, vibrate with very different degrees of intensity, is easily demonstrated. Suspending a little ball of sealing wax *a*, fig. 75, by a string, and allowing it to rest gently against the interior surface of this inverted bell, it is tossed to and fro when the bell is thrown into vibration. But the rattling of the sealing wax ball is far more violent when it rests against the vibrating segments than when it rests against the nodes. Permitting the ivory bob of a short pendulum to rest in succession against a vibrating segment and against a node of the Great Bell of Westminster, I found that in the former position it was driven away five inches, in the latter only two inches and three-quarters when the hammer fell upon the bell.

Could the 'Great Bell' be turned upside down and filled with water, on striking it the vibrations would express themselves in beautiful ripples upon the liquid surface. Similar ripples may be obtained with smaller bells, or even with finger and claret glasses, but they would be too minute for my present purpose. I have here a large hemispherical glass which emits a full deep note. I fill it with water and pass the fiddle-bow across its edge; crispations immediately cover its surface. When I draw the bow vigorously, you see the water rising in a copious spray of liquid spherules from the four vibrating segments. I will endeavour to show you these sonorous ripples. The broad beam from the electric lamp being permitted to fall upon the tranquil water is now reflected at the proper angle, and in the path of the reflected beam I place this large lens, which throws a magnified image of the water surface upon the screen. I now pass the bow gently across the edge of the glass, or I rub my finger gently along the edge; you hear this low sound, and at the same time you observe the ripples breaking in visible music over the four sectors of the liquid surface.*

When bisulphide of carbon is employed instead of water, its spherules, in consequence of the greater weight of the liquid, bound from it with greater momentum, and the exquisite mosaic upon its surface is longer retained. But a more beautiful effect is produced when one of the lighter volatile liquids is made use of. You know the experiment of Leidenfrost which illustrates the spheroidal condition of water. You know that if water be poured into a red-hot silver basin, instead of flashing at once into steam, it rolls about upon its own vapour. The same effect is produced if we drop a volatile liquid, like ether, on the surface of warm water. The drop retains its spheroidal shape. Filling a bell-glass with ether or alcohol, a sharp sweep of the bow over the edge of the glass detaches the liquid spherules, which, when they fall back, do not mix

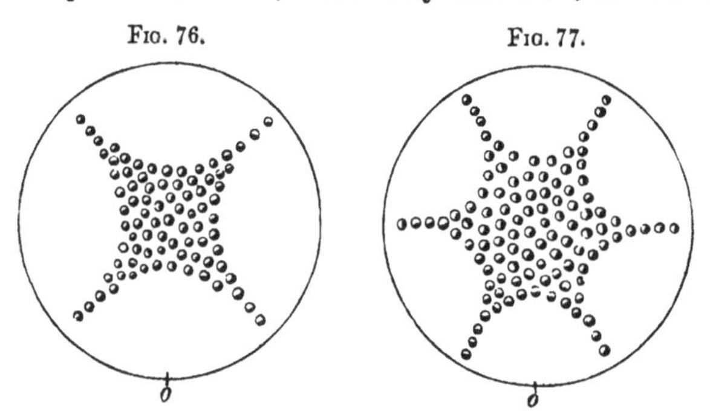

FIG. 76. FIG. 77.

with the liquid, but are driven over the surface on wheels of vapour to the nodal lines. The warming of the liquid, as might be expected, improves the effect. M. Melde, to whom we are indebted for this beautiful experiment, has given the drawings, figs. 76 and 77, representing what

* For the illustration of this subject, I am indebted to the kindness of Mr. Bird, of Birmingham, who executed several airs on his magnificent set of bell glasses, which had been sent to London for this express purpose.

occurs when the surface is divided into four and into six vibrating parts. With a thin wine glass and strong brandy the effect may also be obtained.

The glass and the liquid within it vibrate here together, and everything that interferes with the perfect continuity of the entire mass disturbs the sonorous effect. A crack in the glass passing from the edge downwards would extinguish its sounding power. The same effect is produced by a rupture in the continuity of the liquid. To demonstrate this, I have placed in this glass a solution of carbonate of soda. I strike the glass, and you hear this clear musical sound. But I now add a little tartaric acid to the liquid; it foams, and this dry unmusical collision takes the place of the musical sound. As the foam disappears the sonorous power returns, and now that the liquid is once more clear, you hear the musical ring as before.

Fig. 78.

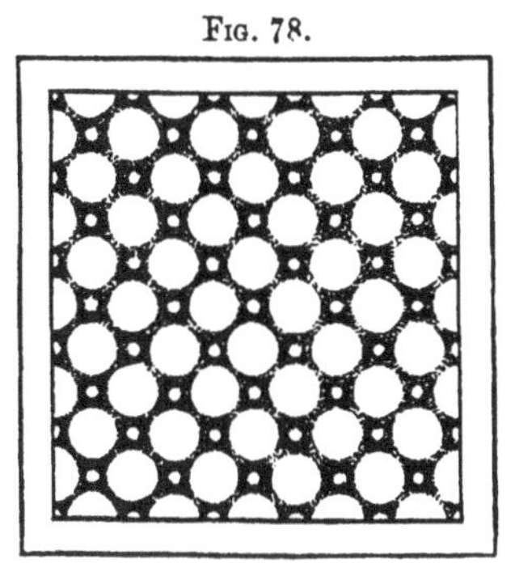

The ripples of the tide leave their impressions upon the sand over which they pass. The ripples produced by sonorous vibrations have been proved by Mr. Faraday competent to do the same. Attaching a plate of glass to a long flexible board, and pouring a thin layer of water over the surface of the glass, on causing the board to vibrate, its tremors chase the water into a beautiful mosaic of ripples. A thin stratum of sand strewn upon the plate, is acted upon by the water, and carved into patterns, of which fig. 78 is a reduced specimen.

SUMMARY OF LECTURE IV.

A rod fixed at both ends and caused to vibrate transversely divides itself in the same manner as a string vibrating transversely.

But the succession of its overtones is not the same as that of a string, for while the series of tones emitted by the string is expressed by the natural numbers 1, 2, 3, 4, 5, &c.; the series of tones emitted by the rod is expressed by the squares of the odd numbers 3, 5, 7, 9, &c.

A rod fixed at one end can also vibrate as a whole, or it can divide itself into vibrating segments separated from each other by nodes.

In this case the rate of vibration of the fundamental tone is to that of the first overtone as 4 : 25, or as the square of 2 to the square of 5. From the first division onwards the rates of vibration are proportional to the squares of the odd numbers 3, 5, 7, 9, &c.

With rods of different lengths the rate of vibration is inversely proportional to the square of the length of the rod.

Attaching a glass bead silvered within to the free end of the rod, and illuminating the bead, the spot of light reflected from it describes curves of various forms when the rod vibrates. The Kaleidophone of Wheatstone is thus constructed.

The iron fiddle and the musical box are instruments, whose tones are produced by rods, or tongues, fixed at one end and free at the other.

A rod free at both ends can also be rendered a source of sonorous vibrations. In its simplest mode of division it has two nodes, the subsequent overtones correspond to divisions by 3, 4, 5, &c. nodes. Beginning with its first mode of division the tones of such a rod are represented by the squares of the odd numbers 3, 5, 7, 9, &c.

The claque-bois, straw-fiddle, and glass harmonica are instruments whose tones are those of rods or bars free at both ends, and supported at their nodes.

When a straight bar, free at both ends, is gradually bent at its centre, the two nodes corresponding to its fundamental tone gradually approach each other. It finally assumes the shape of a tuning-fork which, when it sounds its fundamental note, is divided by two nodes near the base of its two prongs into three vibrating parts.

There is no division of a tuning-fork by three nodes.

In its second mode of division, which corresponds to the first overtone of the fork, there is a node on each prong and two others at the bottom of the fork.

The fundamental tone of the fork is to its first overtone approximately as the square of 2 is to the square of 5. The vibrations of the first overtone are, therefore, about $6\frac{1}{4}$ times as rapid as those of the fundamental. From the first overtone onwards the successive rates of vibration are as the squares of the odd numbers 3, 5, 7, 9, &c.

We are indebted to Chladni for the experimental investigation of all these points. He was enabled to conduct his enquiries through the discovery that, when sand is scattered ver a vibrating surface, it is driven from the vibrating portions of the surface, and collects along the nodal lines.

Chladni embraced in his investigations plates of various forms. A square plate, for example, clamped at the centre, and caused to emit its fundamental tone, divides itself into four smaller squares by lines parallel to its sides.

The same plate can divide itself into four triangular vibrating parts, the nodal lines coinciding with the diagonals. The note produced in this case is a fifth above the fundamental note of the plate.

The plate may be further subdivided, sand-figures of extreme beauty being produced; the notes rise in pitch as the subdivision of the plate becomes more minute.

These figures may be deduced from the coalescence of different systems of vibration.

When a circular plate clamped at its centre sounds its fundamental tone, it divides into four vibrating parts, separated by four radial nodal lines.

The next note of the plate corresponds to a division into six vibrating sectors, the next note to a division into eight sectors; such a plate can divide into any even number of vibrating sectors, the sand-figures assuming beautiful stellar forms.

The rates of vibration corresponding to the divisions of a disc are represented by the squares of the numbers 2, 3, 4, 5, 6, &c. In other words, the rates of vibration are proportional to the squares of the numbers representing the sectors into which the disc is divided.

When a bell sounds its deepest note it is divided into four vibrating parts separated from each other by nodal lines, which run upwards from the sound-bow and cross each other at the crown.

It is capable of the same subdivision as a disc: the succession of its tones being also the same.

LECTURE V.

LONGITUDINAL VIBRATIONS OF A WIRE—RELATIVE VELOCITIES OF SOUND IN BRASS AND IRON—LONGITUDINAL VIBRATIONS OF RODS FIXED AT ONE END—OF RODS FREE AT BOTH ENDS — DIVISIONS AND OVERTONES OF RODS VIBRATING LONGITUDINALLY — EXAMINATION OF VIBRATING BARS BY POLARISED LIGHT—DETERMINATION OF VELOCITY IN SOLIDS—RESONANCE—VIBRATIONS OF STOPPED PIPES: THEIR DIVISIONS AND OVERTONES —RELATION OF THE TONES OF STOPPED PIPES TO THOSE OF OPEN PIPES—CONDITION OF COLUMN OF AIR WITHIN A SOUNDING ORGAN-PIPE—REEDS AND REED-PIPES — THE ORGAN OF VOICE—OVERTONES OF THE VOCAL CHORDS—THE VOWEL SOUNDS—KUNDT'S EXPERIMENTS—NEW METHODS OF DETERMINING VELOCITY OF SOUND.

WE have thus far occupied ourselves exclusively with transversal vibrations; that is to say, vibrations executed at right angles to the lengths of the strings, rods, plates, and bells subjected to examination. A string is also capable of vibrating in the direction of its length, but here the power which enables it to vibrate is not a tension applied externally, but the elastic force of its own molecules. Now this molecular elasticity is much greater than any that we can ordinarily develope by stretching the string, and the consequence is that the sounds produced by the *longitudinal vibrations* of a string are, as a general rule, much more acute than those produced by its transverse vibrations. These longitudinal vibrations may be excited by the oblique passage of a fiddle-bow; but they are more easily produced by passing briskly along the string a bit of cloth or leather on which powdered resin has been strewn The resined fingers answer the same purpose.

I pluck the wire of our monochord thus aside; you hear the sound produced by its transversal vibrations. I now rub this resined leather along the wire; a note much more piercing than the last is heard, which is due to the longitudinal vibrations of the wire. Behind the table is stretched a stout iron wire, 23 feet long. One end of it is firmly attached to an immovable wooden tray, the other end is coiled round a pin fixed firmly into one of our benches. With a key this pin can be turned, and the wire stretched so as to facilitate the passage of the rubber. I clasp the wire with the resined leather, and pass my hand to and fro along it; a rich loud musical sound is audible to you all. I tightly clip the wire at its centre, and rub one of its halves; the note is now the octave of what it was before, the vibrations being twice as rapid. I clip the wire at one-third of its length and rub the shorter segment; this note is a fifth above the octave. I clip it at one-fourth of its length and rub that fourth; this note is the double octave of that yielded by the whole wire, being produced by four times the number of vibrations. Thus, in longitudinal as well as in transversal vibrations, the number of vibrations executed in a given time is inversely proportional to the length of the wire.

And notice the surprising power of these sounds when the wire is rubbed vigorously. I take a shorter length, the note rises; shorter still, and now it is so acute, and at the same time so powerful, as to be hardly bearable. It is not the wire itself which produces this intense sound; it is the wooden tray at its end to which its vibrations are communicated. And the vibrations of the wire being longitudinal, those of the tray, which is at right angles to the wire, must be transversal. We have here, indeed, an instructive example of the conversion of longitudinal into transverse vibrations.

Abandoning the bridge by which I clipped the wire, I cause it again to vibrate longitudinally through its entire length. While I do so my assistant will turn the key at the end, and thus change the tension. You notice no variation of the note. Once sufficiently stretched to enable it to abut against its points of attachment, the longitudinal vibrations of the wire, unlike the transverse ones, are independent of the tension. Observe now that I have here a second wire of brass of the same length and thickness as the iron one. I rub them both. Their tones are not the same; that of the iron wire is considerably the higher of the two. Why? Simply because the velocity of the sound-pulse is greater in iron than in brass. The pulses in this case pass to and fro from end to end of the wire. At one moment the wire pushes against the tray at its end. At the next moment the wire pulls the tray, this pushing and pulling being due to the passage of the pulse to and fro along the whole wire. The time required for a pulse to run from one end to the other *and back* is that of a complete vibration. In that time the wire imparts one push and one pull to the wooden tray at its end; the wooden tray imparts one complete vibration to the air, and the air bends once in and once out the tympanic membrane. Hence it is manifest that the rapidity of vibration, or, in other words, the pitch of the note, depends upon the velocity with which the sound-pulse is transmitted through the wire.

And now the solution of a beautiful problem falls of itself into our hands. Without quitting this room, we can determine the relative velocities of sound through brass and iron. I shorten the brass wire until the note emitted by it is of the same pitch as that emitted by its fellow. You hear both notes now sounding in unison. This proves that the sound-pulse travels through these 23 feet of iron wire, and through these 15 feet 6 inches of brass wire in the same time. These lengths are in the ratio of 11 : 17, and these two numbers express the relative velocities of sound in brass and iron. In fact, the former velocity is 11,000 feet, and the latter 17,000 feet a second. The same method is of course applicable to many other metals.

When a wire, vibrating longitudinally, emits its lowest note, there is no node whatever upon it: the pulse, as I have said, runs to and fro along the whole length of the wire. But, like a string vibrating transversely, it can also subdivide itself into ventral segments separated by nodes. By damping the centre of the wire we make that point a node. The pulses here run from both ends, meet in the centre, recoil from each other, and return to the

Fig. 79.

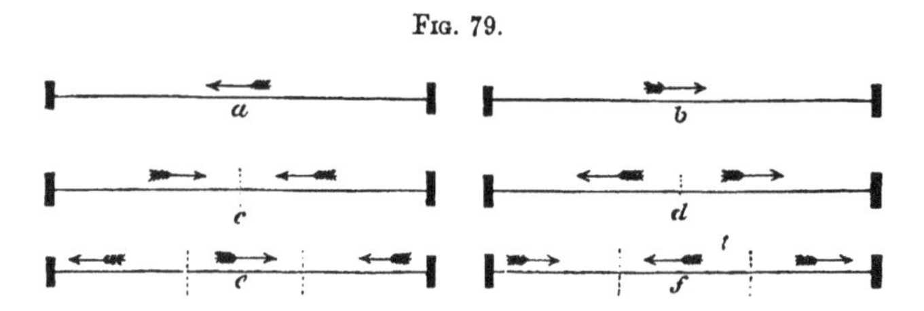

ends, where they are reflected as before. The note produced by the wire divided thus into two vibrating segments is the octave of its fundamental note. The next higher note is produced by the division of the wire into three vibrating segments, separated from each other by two nodes. The first of these three modes of vibration is shown in fig. 79, *a* and *b*; the second at *c* and *d*; the third at *e* and *f*; the nodes being marked by dotted transverse lines, and the arrows in each case pointing out the direction in which the pulse moves. The tones of the wire, as determined by their rates of vibration, follow the order of the numbers 1, 2, 3, 4, 5, &c., just as in the case of a wire vibrating transversely.

A *rod* or *bar* of wood or metal, with its two ends fixed, and vibrating longitudinally, divides itself in the same manner as the wire. The succession of tones is also the same in both cases.

Rods and bars *with one end fixed* are also capable of vibrating longitudinally. A smooth wooden or metal rod for example, with one of its ends fixed in a vice, yields a musical note, when the resined fingers are passed along it. When such a rod yields its lowest note, it simply elongates and shortens in quick alternation. There is, then, no node upon the rod. When rods of different lengths are compared, the pitch of the note is inversely proportional to the length of the rod. This follows necessarily from the fact, that the time of a complete vibration is the time required for the sonorous pulse to run twice to and fro over the rod. The first overtone of a rod, fixed at one end, corresponds to its division by a node at a point one-third of its length from its free end. Its second overtone corresponds to a division by two nodes, the highest of which is at a point one-fifth of the length of the rod from its free end, the remainder of the rod being divided into two equal parts by the second node. In fig. 80, *a* and *b*, *c* and *d*, *e* and *f*, are shown the conditions of the rod answering to its first three modes of vibration: the nodes as before, are marked by dotted lines, the arrows in the respective cases marking the direction of the pulses.

FIG. 80.

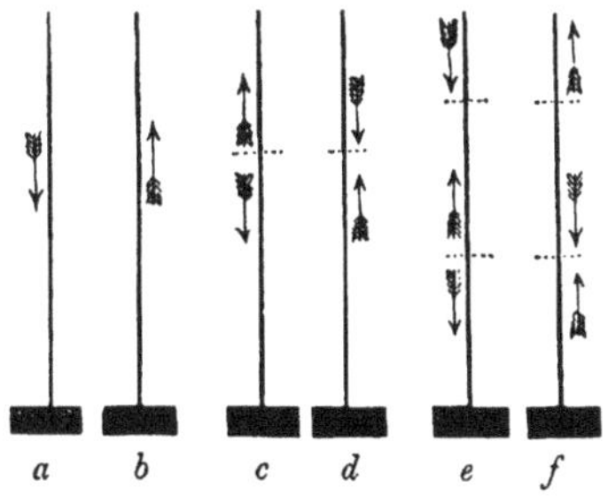

a *b* *c* *d* *e* *f*

The order of the tones of a rod fixed at one end and vibrating longitudinally is that of the odd numbers 1, 3, 5, 7, &c. It is easy to see that this must be the case. For the time of vibration of *c* or *d* is that of the segment above the dotted line: and the length of this segment being only one-third that of the whole rod, its vibrations must be three times as rapid. The line of vibration in *e* or *f* is also that of its highest segment, and as this segment is one-fifth of the length of the whole rod, its vibrations must be five times as rapid. Thus the order of the tones must be that of the odd numbers.

Before you, fig. 81, is a musical instrument, the sounds of which are due to the longitudinal vibrations of a number of deal rods of different lengths. Passing the resined fingers over the rods in succession, I obtain a series of notes of varying pitch. It would, however, require a far more expert performer than I am to render the tones of this instrument pleasant to you.

FIG. 81.

Rods *with both ends free* are also capable of vibrating longitudinally, and producing musical tones. The investigation of this subject will lead us to exceedingly important results. Clasping a long glass tube exactly at its centre with my left hand, I pass with my right a wetted cloth over one of its halves; a clear musical sound is the result. A solid glass rod of the same length would yield the same note. In this case the centre of the tube is a node, and the two halves elongate and shorten in quick alternation.

M. König of Paris has provided me with an instrument which will illustrate this action. This rod of brass, *a b*, fig. 82, is held at its centre by the clamp, *s*, while an ivory ball, suspended by two strings from the points, *m* and *n*, of this wooden frame, is caused to rest against the end, *b*, of the brass rod. Drawing gently a bit of resined leather over the rod near *a*, it is thrown into longitudinal

FIG. 82.

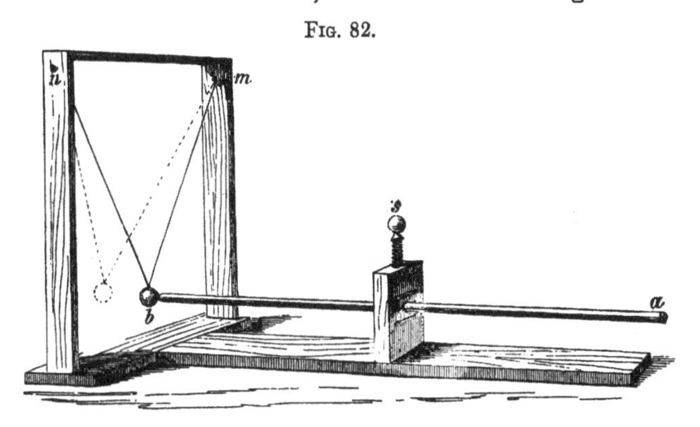

vibration. The centre, *s*, is a node; but the uneasiness of the ivory ball shows you that the end, *b*, is in a state of tremor. I apply the rubber still more briskly. The ball, *b*, rattles, and now the vibration is so intense, that the ball is rejected with violence whenever it comes into contact with the end of the rod.

When I pass the wetted cloth over the surface of this glass tube I can see the film of liquid left behind by the cloth forming narrow tremulous rings all along the rod. Now this shivering of the liquid is due to the shivering of the glass underneath it, and it is possible so to augment the intensity of the vibration that the glass shall actually go to pieces. Savart was the first to show this. Twice in this place I have repeated his experiment, sacrificing in each case a fine glass tube 6 feet long and 2 inches in diameter. Seizing the tube at its centre, c, fig. 83,

I swept my hand vigorously to and fro along C D, until finally the half most distant from my hand was shivered into annular fragments. On examining these it was found that, narrow as they were, many of them were marked by circular cracks indicating a still more minute subdivision.

Fig. 83.

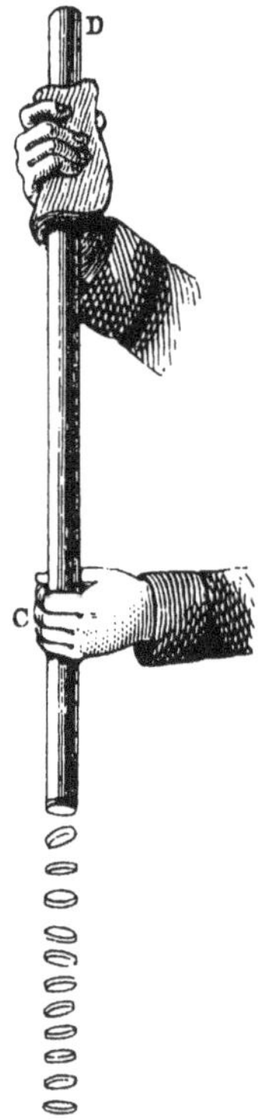

In this case also the rapidity of vibration is inversely proportional to the length of the rod. A rod of half the length vibrates longitudinally with double the rapidity, a rod of one-third the length with treble the rapidity, and so on. The time of a complete vibration being that required by the pulse to travel to and fro over the rod, and that time being directly proportional to the length of the rod, the rapidity of vibration must, of necessity, be in the inverse proportion.

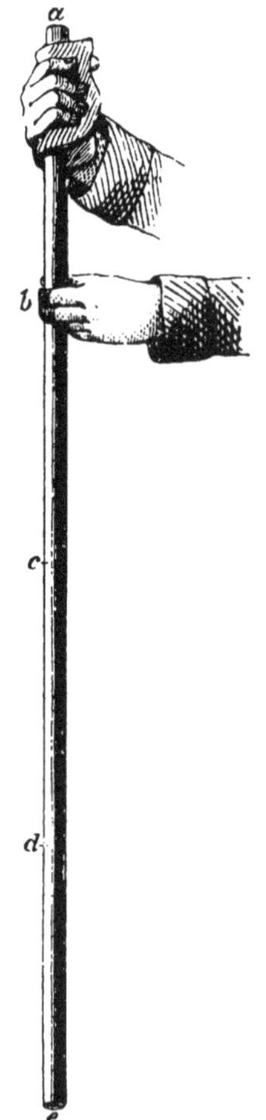

Fig. 84.

This division of a rod by a single node at its centre corresponds to the deepest tone produced by its longitudinal vibration. But, as in all other cases hitherto examined, such rods can subdivide themselves further. I lay hold of the long glass rod, *a e*, fig. 84, at a point *b*, midway between its centre and one of its ends, and rub its short section, *a b*, with a wet cloth. The point *b* becomes a node, and a second node, *d*, is formed at the same distance from the opposite end of the rod. Thus we have the rod divided into three vibrating parts, consisting of one whole ventral segment, *b d*, and two half ones, *a b* and *d e*. The sound corresponding to this division of the rod is the octave of its fundamental note.

You have now a means of bringing two of my statements face to face, and of proving whether they do or do not contradict each other. For, if the second mode of division just described produces the octave of the fundamental note, and if a rod of half the length produces the same octave, then the whole rod held at a point one-fourth of its length from one of its ends ought to emit the same note as the half rod held in the middle. A second person shall sound the half rod, while I sound the whole one; you hear both sounds, and in pitch they are identical.

FIG. 85.

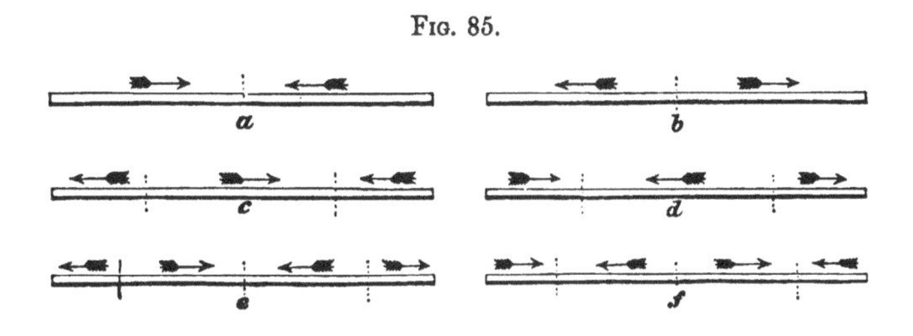

In fig. 85, *a* and *b*, *c* and *d*, *e* and *f*, are shown the three first divisions of a rod free at both ends and vibrating longitudinally. The nodes as before are marked by transverse dots, and the direction of the pulses shown by arrows. The order of the tones is that of the numbers 1, 2, 3, 4, &c.

When a tube or rod vibrating longitudinally yields its fundamental tone, its two ends are in a state of free vibration, the glass there suffering neither strain nor pressure. The case at the centre is exactly the reverse; here there is no vibration, but a quick alternation of tension and compression. When the sonorous pulses meet at the centre they squeeze the glass; when they recoil they strain it. Thus while at the ends we have the maximum of vibration, but no change of density, at the centre we have the maximum changes of density, but no vibration.

We have now cleared the way for the introduction of a very beautiful experiment made many years ago by Biot, but never, as far as I am aware of, repeated on the scale here presented to you. I send the light from our electric lamp, L, fig. 86, through a prism, B, of bi-refracting spar, and thus obtain a beam of polarised light. This beam impinges on a second prism of spar, *n*, but, though both prisms are perfectly transparent, the light which has passed through the first cannot get through the second. And now I wish to show you that, by introducing a piece of glass between the two prisms, and subjecting the glass to either strain or pressure, the light will be enabled to pass through the entire system. I operate with this rectangular glass rod six inches long. In its natural condition you observe it is inactive; when the rod is placed between the prisms, the screen remains dark. By the mere force of my fingers I bend the glass rod, thereby throwing one of its halves into a state of strain, and the other half into a state of compression, and now you see the illuminated image of the rod upon the screen. Along the centre of the rod you observe a band of darkness: this band marks the place of transition from strain to pressure, where the glass is in its natural condition, and therefore unable to affect the light.

Abandoning the glass rod, I now introduce between the

prisms B and n, a rectangle, $s\ s'$, of plate-glass, 6 feet long, 2 inches wide, and $\frac{1}{3}$ of an inch thick, which I intend to throw into longitudinal vibration. The beam from L passes through the glass at a point near its centre, which is held in a vice, c, so that when a wet cloth is passed over one of the halves, $c\ s$, of the strip, the centre will be a node.

Fig. 86.

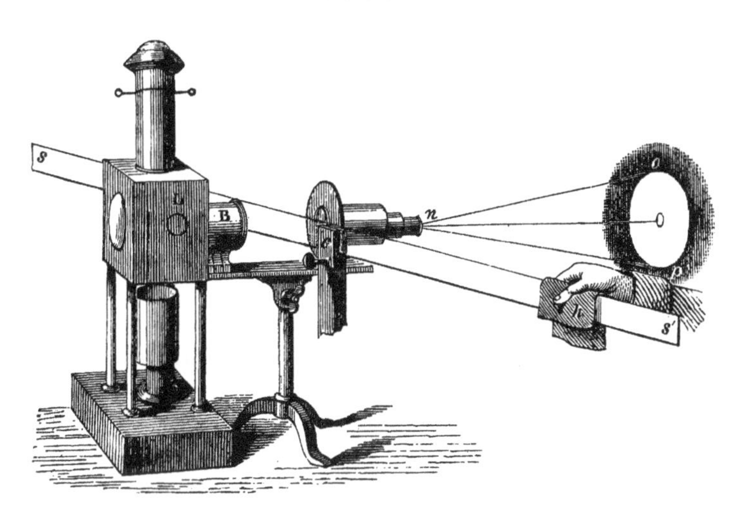

During its longitudinal vibration the glass near the centre is, as already explained, alternately strained and compressed; and this successive strain and pressure confers upon the glass a structure which enables it to abolish the particular condition of the light on which its extinguishability by the second prism, n, depends. Watch the experiment. The screen is all dark: I pass my wetted cloth briskly over the glass; you hear its sound, and at the same time a brilliant disc of light, three feet in diameter, flashes out upon the screen. The vibration quickly subsides, and the luminous disc as quickly disappears, to be, however, revived at will by the passage of the wetted cloth along the glass.

The light of this disc appears to be continuous, but it is really intermittent, for it is only when the glass is under strain, or pressure, that the light can get through. In passing from strain to pressure, and from pressure to strain, the glass is for a moment in its natural state, which, if it continued, would leave the screen dark. But the impressions of brightness, due to the strains and pressures, remain far longer upon the retina than is necessary to abolish the intervals of darkness. Hence the screen appears illuminated by a continuous light. I now arrange the two prisms so that the beam in the first instance passes through both ; it might be expected that darkness will now result when the glass is caused to vibrate. This, however, is not the case, though dark intervals undoubtedly occur. You notice a diminution of the light, but you have no extinction. The dark intervals, in fact, are of such short duration as to be almost abolished by the intervals of light which precede and succeed them.

And now I shift the rectangle of glass so as to cause the beam of polarised light to pass through it close to its end, *s*. The longitudinal vibrations of the glass have now no effect whatever upon the polarised beam. Thus, by means of this subtle investigator, we demonstrate that while the centre of the glass, where the vibration is nil, is subjected to quick alternations of strain and pressure, the ends of the rectangle, where the vibration is a maximum, suffer neither.*

Thus far I have operated almost exclusively with glass rods and tubes; but I have also rods of wood and metal, which yield musical tones when they vibrate longitudinally. Here, however, the rubber employed is not a wet cloth, but

* This experiment succeeds almost equally well with a glass *tube*.

a piece of leather, on which powdered resin has been strewn. The resined fingers also elicit the music of the rods. Their modes of vibration are those already indicated, the pitch, however, varying with the velocity of the sonorous pulse in the respective substances. I have here two rods of the same length, the one of deal, the other of Spanish mahogany; let us sound them together. The pitch of the one is much lower than that of the other. Why? simply because the sonorous pulses pass more slowly through this particular kind of mahogany than through deal. Can we find the relative velocity of sound through these bodies? With the greatest ease. We have only to shorten the mahogany rod till it yields the same note as the deal one. This I do carefully and by degrees. The notes, rendered approximate by the first trials, are now identical. Through this rod of mahogany 4 feet long, and through this rod of deal 6 feet long, the sound-pulse passes in the same time, and these numbers express the relative velocities of sound through the two substances.

Modes of investigation, which I could do no more than hint at in our earlier lectures, are now falling naturally into our hands. When in my first lecture I spoke of the velocity of sound in air, no doubt many possible methods of determining this velocity occurred to your minds, because here you had miles of space to operate upon. But how are we to determine the velocity of sound through wood or metal where such distances are out of the question? In the simple manner just indicated: from the notes which they emit when suitably prepared, we may infer with certainty the *relative* velocities of sound through different solid substances; and determining the ratio of the velocity in any one of them to its velocity in air, we are able to draw up a table of absolute velocities. But how is air to be introduced into the series? We shall soon be able to answer this question, approaching it, however, through a number of phenomena with which, at first sight, it appears to have no connection.

RESONANCE.

A series of tuning-forks stands before you, whose rates of vibration have already been determined by the syren. This one, you will remember, vibrates 256 times in a second, the length of the sonorous wave which it produces being, therefore, 4 feet 4 inches. The fork is now detached from its case, so that when struck against its pad you hardly hear it. I hold the vibrating fork over this glass jar,

Fig. 87.

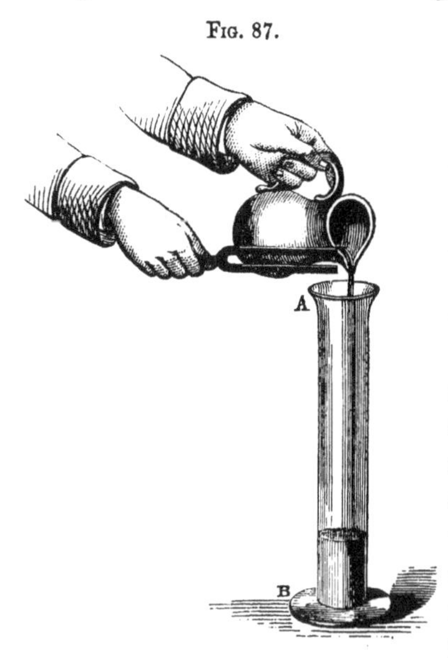

A B, fig. 87, 18 inches deep; but you still fail to hear the sound of the fork. Preserving the fork in its position, I pour water with the least possible noise into the jar. The column of air underneath the fork becomes shorter as the water rises. The sound, you observe, augments in intensity; and when the water has reached a certain level it bursts forth with extraordinary power. I continue to pour in water, the sound sinks, and becomes finally as inaudible as at first. By pouring the water carefully out, I reach a point where the reinforcement of the sound again occurs. Experimenting thus, I learn that there is one particular length of the column of air which, when the fork is placed above it, produces a maximum augmentation of the sound. This reinforcement of the sound is named *resonance.*

Operating in the same way with all the forks in succession, I find for each a column of air which yields a maximum resonance. These columns are of different lengths, becoming shorter as the rapidity of vibration increases. In fig. 88, the series of jars is represented, the number of vibrations to which each resounds being placed above it.

Fig. 88.

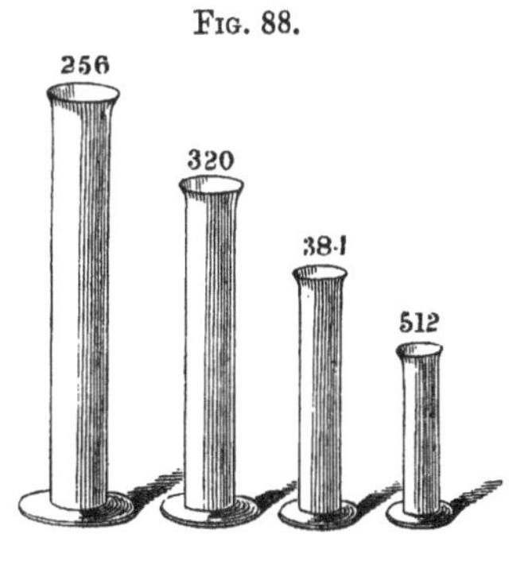

But why should the jars shorten, and what is the physical meaning of this very wonderful effect? The greater volume of sound heard everywhere throughout the room can only be due to the greater amount of motion communicated to the air of the room. Under what circumstances is the fork enabled to communicate this increase of motion? To solve this question we must revive our knowledge of the relation of the motion of the fork to the motion of the sonorous wave produced by the fork. Supposing a prong of this fork, which executes 256 vibrations in a second, to vibrate between the points *a* and *b*, fig. 89. In its motion from *a* to *b* the fork generates half a sonorous wave, and as the length of the whole wave emitted by this fork is 4 feet 4 inches, at the moment the prong reaches *b*, the foremost point of the sonorous wave will be at c, 2 feet 2 inches distant from the fork. The motion of the wave, then, is vastly greater than that of the fork. In fact the distance *a b* is, in this case, not more than one-twentieth of an inch, and in the time required by the fork to describe this distance, the wave has passed over a distance of 26 inches. With forks of lower pitch, the difference would be still greater.

FIG. 89.

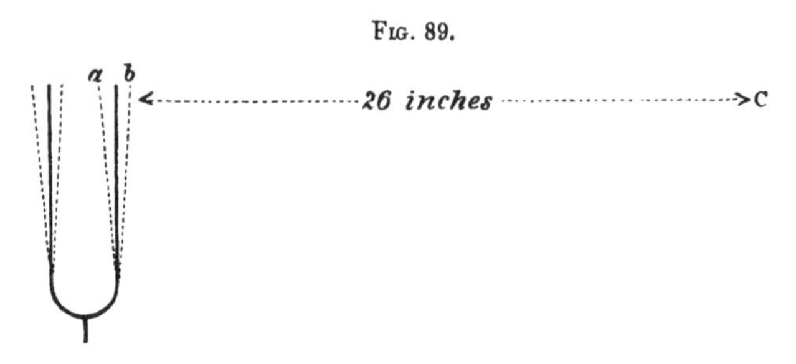

Our next question is, what is the length of the column of air which most powerfully resounds to this fork? By measurement with a two-foot rule I find it to be 13 inches. But the length of the wave emitted by the fork is 52 inches; hence *the length of the column of air which resounds to the fork is equal to one-fourth of the length of the wave produced by the fork.* This rule is general, and might be illustrated by any other of the forks instead of this one.

FIG. 90.

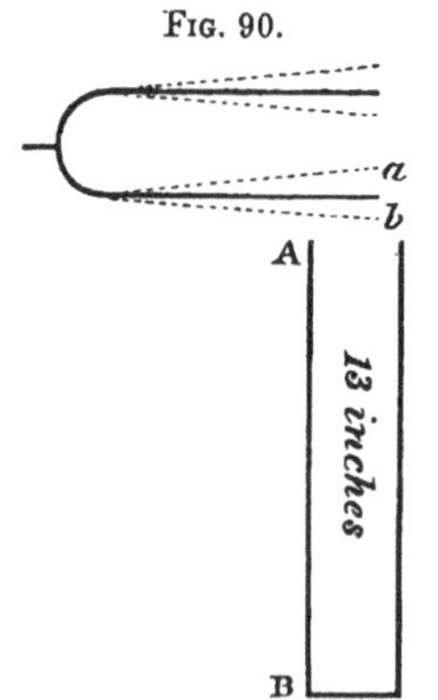

Figure, then, to your minds the prong, vibrating between the limits *a* and *b*, placed over its resonant jar, A B, fig. 90. In the time required by the fork to move from *a* to *b*, the condensation which it produces runs down to the bottom of the jar, is there reflected, and, as the distance to the bottom and back is 26 inches, the reflected wave will reach the fork at the moment when it is on the point of returning from *b* to *a*. The rarefaction of the wave is produced by the retreat of the prong from *b* to *a*. This rarefaction will also run to the bottom of the jar and back, overtaking the prong just as it reaches the limit, *a*, of its excursion. It is plain from this analysis, that the vibrations of the fork are perfectly synchronous with the vibrations of the aërial column, AB; and in virtue of this synchronism the motion accumulates in the jar, spreads abroad in the room, and produces this vast augmentation of the sound.

Supposing we substitute for the air in one of these jars a gas of different elasticity, we should find the length of the resounding column to be different. The velocity of sound through coal-gas is to its velocity in air about as 8 : 5. Hence, to synchronise with our fork, a jar filled with coal-gas would have to be deeper than a jar filled with air. I turn this jar, 18 inches long, upside down, and hold it over the end of this closed gas orifice. I agitate our tuning-fork, and hold it close to the mouth of the jar: it is scarcely audible. The jar, with air in it, is 5 inches too deep for this fork. I now turn on the gas, and as it ascends in the jar the note swells out, proving that for the more elastic gas a depth of 18 inches is not too great. In fact it is not great enough; for if I allow too much gas to enter, the resonance is weakened. By suddenly turning the jar upright, still holding the fork close to its mouth, the gas escapes, and at the point of proper admixture of gas and air the note swells out again.*

We are indebted to Savart for the illustration of the effect of resonance which I have now to bring before you. I have here a fine sonorous bell, fig. 91, which I throw into intense vibration by passing a resined bow across its edge. You hear its sound, pure but not very forcible. I cause

* The escape of the gas from the jar is so very rapid as to lessen the effect of this experiment. It is more easily executed with hydrogen than with coal-gas.

the open mouth of this large tube, which is closed at one end, to approach one of the vibrating segments of the bell. You notice an augmentation of the sound: and now that the tube is close to the bell, the tone swells into a roar of exceeding power, but perfectly smooth and musical. Thus, as I alternately withdraw and advance the tube, the

Fig. 91.

sound sinks and swells in this extraordinary manner. I allow it to sink until the unaided bell becomes perfectly inaudible. I move the tube forward; once more the sound, which a moment ago was unheard by those nearest to the bell, is now audible throughout the entire room. I now take a second tube capable of being lengthened and shortened by a telescopic slider, but, unlike the last, open at both ends. I bring it near the vibrating bell, but the resonance is feeble. I lengthen the tube by drawing out the slider, and, at a certain point, the tone swells out as before. If I make the tube longer, the resonance is again enfeebled. I would draw your attention to the fact, which shall be explained presently, that the open tube which gives the maximum resonance is exactly twice the length of the closed one.

When in the third lecture I took the end of an india-rubber tube in my hand, and found it necessary to time my impulses properly to produce the various ventral segments, I could feel that the quantity of muscular work done, when my impulses were properly timed, was greater than when they were irregular. The same truth may be illustrated by a claret glass half filled with water. Endeavour to move your hand to and fro in accordance with the oscillating period of the water: when you have thoroughly established synchronism, the work thrown upon the hand makes you feel as if the weight of the water had become greater. So likewise with our tuning-fork; when its impulses are timed to the vibrations of the column of air contained in this jar, its work is greater than when they are not so timed. As a consequence of this the tuning-fork comes sooner to rest when it is placed over the jar than when it is permitted to vibrate either in free air, or over a jar of a depth unsuited to its periods of vibration.*

Reflecting on what we have now learned, you would, I am persuaded, have no difficulty in solving the following beautiful problem:—You are provided with a tuning-fork and a syren, and are required by means of these two instruments to determine the velocity of sound in air. To solve this problem, you lack, if anything, the mere power of manipulation which practice imparts. You would propose first to determine, by means of the syren, the number of vibrations executed by your tuning-fork in a second; you would then determine the length of the column of air which resounds to the fork. This length multiplied by 4 would give you, approximately, the length of the sonorous wave, and the wave-length multiplied by the number of vibrations in a second would give you the required

* Only an extremely small fraction of the fork's motion is converted into sound. The remainder is expended in overcoming the internal friction of its own particles. In other words, nearly the whole of the motion is converted into heat.

velocity. Without quitting your private room, therefore, you could solve this important problem. We will go on, if you please, in this fashion, making our footing sure as we advance.

ORGAN-PIPES.

Before me on the table are two resonant jars, and in my two hands I hold two tuning-forks. I agitate both forks, and hold both of them over this jar. One of them only is heard. I hold them over the other jar; the other fork alone is now heard. Each jar selects that fork for reinforcement whose periods of vibration synchronise with its own. Instead of two forks I might hold two dozen over either of these jars; from the confused assemblage of pulses thus generated, the jar would select, and reinforce, that one which corresponds to its own period of vibration.

Fig. 92.

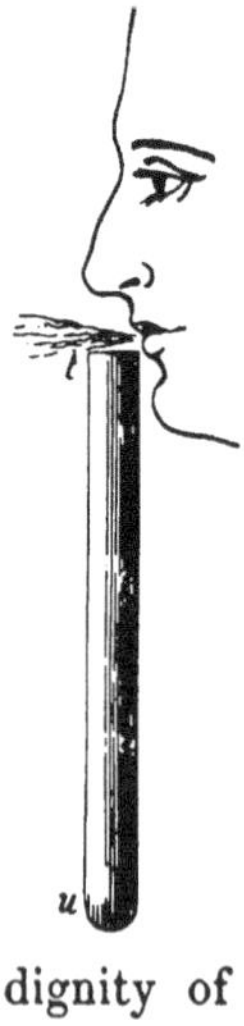

I now take this same jar in my hand, and raising it to the level of my lips, blow *across* its open mouth. Or, better still, for the jar is too wide for this experiment, I blow across the open end of a glass tube, $t\,u$, fig. 92, of the same length as the jar, and $\frac{3}{4}$ths of an inch in diameter. I thereby produce a fluttering of the air; I generate, in fact, an assemblage of pulses at the open mouth of the tube; and what is the consequence? The tube selects that pulse of the flutter which is in synchronism with itself, and raises it to the dignity of a musical sound. The sound, you perceive, is precisely that obtained when the proper tuning-fork is placed over the tube. The column of air within the tube has in this case virtually created its own tuning-fork; for by the reaction of its pulses upon the sheet of air issuing from my lips, it has compelled that sheet to vibrate in synchronism with itself, and made it thus act the part of the tuning-fork.

I pass on to our other tuning-forks, selecting for each of them a resonant tube. In every case, on blowing across the open end of the tube, I produce a tone identical in pitch with that obtained when the proper tuning-fork is held at that end.

When different tubes are compared, the rate of vibration is found to be inversely proportional to the length of the tube. I have here three tubes, 6, 12, and 24 inches long. I blow gently across the two-foot tube, and bring out its fundamental note; I do the same with the 12-inch tube, and find its note to be the octave of that of the 24-inch. I do the same with the six-inch tube, and find its note to be the octave of that of the 12-inch. It is plain that this must be the case; for the rate of vibration depending on the distance which the pulse has to travel to complete a vibration, if in one case this distance be twice what it is in another, the rate of vibration must be twice as slow. In general terms, the rate of vibration is inversely proportional to the length of the tube through which the pulse passes.

But that the current of air should be thus able to accommodate itself to the requirements of the tube, it must enjoy a certain amount of flexibility. A little reflection will show you that the power of the reflected pulse over the current must depend to some extent on the force of the latter. A stronger current, like a more powerfully stretched string, requires a greater force to deflect it, and when deflected vibrates more quickly. Accordingly, to obtain the fundamental note of this tube, I have to blow very gently across its open end; when this is done, a rich, full, and forcible musical tone is produced. If I blow a little stronger, the sound approaches a mere rustle; I blow more strongly still, and now I obtain a tone of much higher pitch than the fundamental one. This is the first overtone of the tube, to produce which the column of air within it has divided itself into two vibrating parts, with a node between them. I blow still stronger, and obtain a still higher note: the tube is now divided into three vibrating parts, separated from each other by two nodes. Once more I blow with sudden strength, a higher note than any before obtained is the consequence.

Fig. 93.

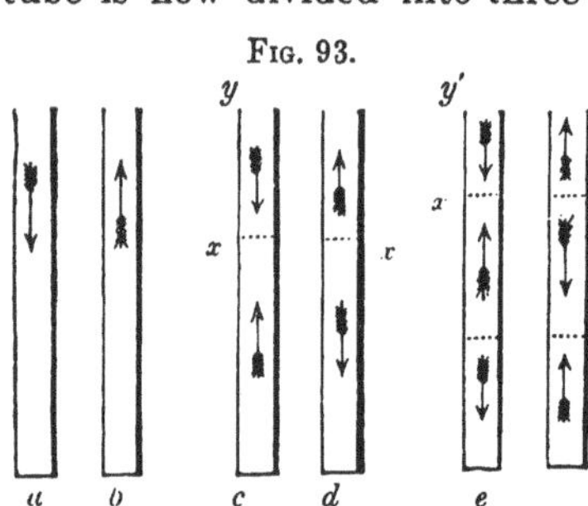

In fig. 93, I have represented the divisions of the column of air corresponding to the first three notes of a tube stopped at one end. At *a* and *b*, which correspond to the fundamental note, the column is undivided; the bottom of the tube is the only node, and the pulse simply moves up and down from top to bottom as denoted by the arrows. In *c* and *d*, which correspond to the first overtone of the tube, we have one nodal surface shown by dots, against which the pulses abut, and from which they are reflected as from a fixed surface. This nodal surface is situated at one-third of the length of the tube from its open end. In *e* and *f*, which correspond to the second overtone, we have two nodal surfaces, the upper one of which is at one-fifth of the length of the tube from its open end, the remaining four-fifths being divided into two equal parts by the second nodal surface. The arrows, as before, mark the directions of the pulses.

We have now to inquire into the relation in which these successive notes stand towards each other. The distance from node to node we have called all through a ventral segment; hence the distance from the middle of a ventral segment to a node is a semi-ventral segment. You will readily bear in mind the law, that *the number of vibrations is directly proportional to the number of semi-ventral segments* into which the column of air within the tube is divided. Thus, when the fundamental note is sounded, we have but a single semi-ventral segment. The bottom here is a node, and the open end of the tube, where the air is agitated, is the middle of a ventral segment. When we obtain the mode of division represented in *c* and *d* we have three semi-ventral segments; in *e* and *f* we have five. The vibrations, therefore, corresponding to this series of notes, augment in the proportion of the series of odd numbers, 1 : 3 : 5. And could we obtain still higher notes, their relative rates of vibration would continue to be represented by the odd numbers, 7, 9, 11, 13, &c. &c.

A moment's consideration will make it evident to you that this must be the law of succession. For the time of vibration in *c* or *d* is that of a stopped tube of the length $x\,y$; but this length is one third of the length of the whole tube, consequently its vibrations must be three times as rapid. The time of vibration in *e* or *f* is that of a stopped tube of the length $x'\,y'$, and inasmuch as this length is one-fifth that of the whole tube, its vibrations must be five times as rapid. We thus obtain the succession 1, 3, 5, and if we pushed matters further we should obtain the continuation of the series of odd numbers.

And here it is once more in your power to subject my statements to an experimental test. I hold in my hands two tubes, one of which is three times the length of the other. I sound the fundamental note of the longest tube, and I then sound the next note above the fundamental. The vibrations of these two notes I have stated to be in the ratio of 1 : 3. This latter note, therefore, ought to be of precisely the same pitch as the fundamental note of the shorter of the two tubes. I sound both tubes: their notes are identical.

I need hardly say that it is only necessary to place a series of these tubes of different lengths thus together, to form that ancient instrument the Pandean pipes, PP′ fig. 94, with which we are all so well acquainted.

FIG. 94.

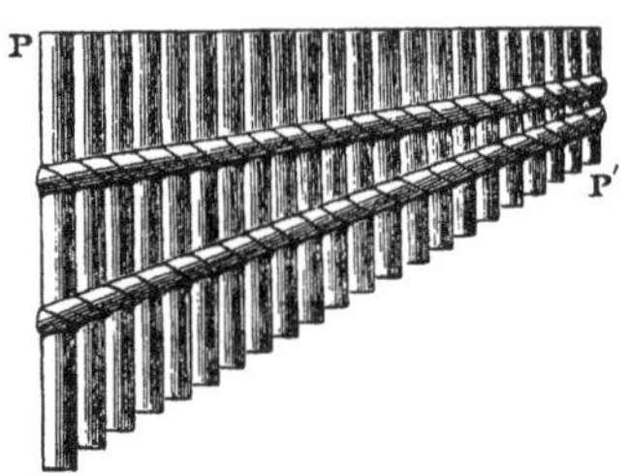

The successive divisions and the relation of the overtones of a rod fixed at one end (described in p. 163), are identical with those of a column of air in a tube stopped at one end, which we have been here considering.

From tubes closed at one end, and which, for the sake of brevity, may be called stopped tubes, we now pass to tubes open at both ends, which we shall call open tubes. Comparing, in the first instance, a stopped tube with an open one of the same length, we find the note of the latter an octave higher than that of the former. This result is general. An open tube always yields the octave of the note of a stopped tube of the same length. From this collection of open tubes I select four, which resound with maximum power to our series of four tuning-forks. Every open tube of the series is double the length of the closed one which resounds to the same fork. To make an open tube yield the same note as a closed one, the former must be twice the length of the latter. And since the length of a closed tube sounding its fundamental note is one-fourth of the length of its sonorous wave, the length of an open tube is one-half that of the sonorous wave that it produces.

It is not easy to obtain a sustained musical note by blowing across the end of one of these open glass tubes; but a

Fig. 95.

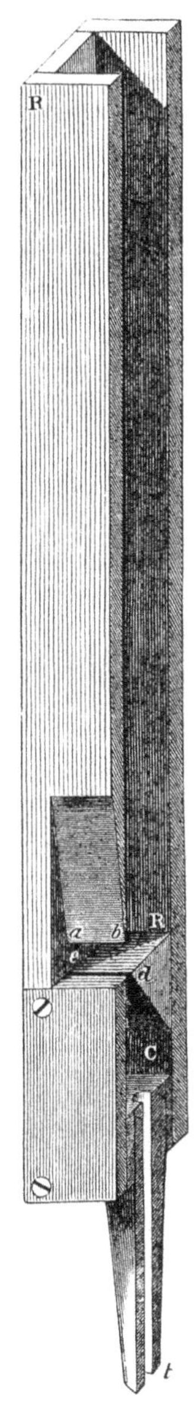

Fig. 96.

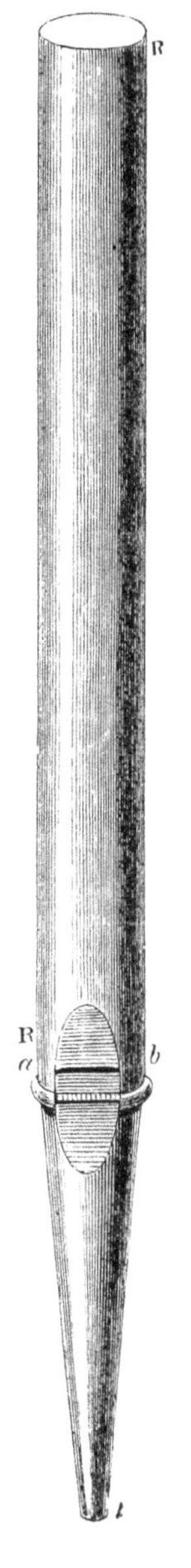

mere puff of breath across the end reveals the pitch to the disciplined ear. In each case it is that of a closed tube half the length of the open one.

There are various ways of agitating the air at the ends of pipes and tubes, so as to throw the columns within them into vibration. In organ-pipes this is done by blowing a thin sheet of air against a sharp edge. This produces a flutter, some particular pulse of which is then converted into a musical sound by the resonance of the associated column of air.

You will have no difficulty in understanding the construction of this open organ-pipe, fig. 95, one side of which has been removed so that you may see its inner parts. Through the tube, *t*, the air passes from the wind-chest into the chamber, c, which is closed at the top, save a narrow slit, *e d*, through which the compressed air of the chamber issues. This thin air current breaks against the sharp edge, *a b*, and there produces a fluttering noise, the proper pulse of which is converted by the resonance of the pipe into a musical sound. The open space between the edge, *a b*, and the slit below it is called the *embouchure*. Fig. 96 represents a stopped pipe of the same length as that shown in fig. 95; and hence producing a note an octave lower.

Instead of causing the air there to flutter, I may place at the embouchure a tuning-fork, whose rate of vibration synchronises with that of the organ pipe, as at A B, fig. 97. The pipe resounds. I take four open pipes of different lengths, and four tuning-forks of different rates of vibration. Beginning with the longest pipe, I strike the most slowly vibrating fork, and bring it near the embouchure. The pipe *speaks* powerfully. I now blow into the same pipe: the tone is identical with that evoked by the tuning-fork. I go through all the pipes in succession, and find in each case that the note yielded by the pipe, when blown into, is exactly that produced when the proper tuning-fork is placed at the embouchure. Conceive all four of these forks placed near the same embouchure; we should have pulses of four different periods there excited; out of the four, however, the pipe would select but one. And if four hundred vibrating forks could be placed there instead of four, the pipe would still make the proper selection. This it also does, when for the pulses of tuning-forks we substitute that assemblage of pulses created by the current of air when it strikes against the sharp upper edge of the embouchure.

Fig. 97.

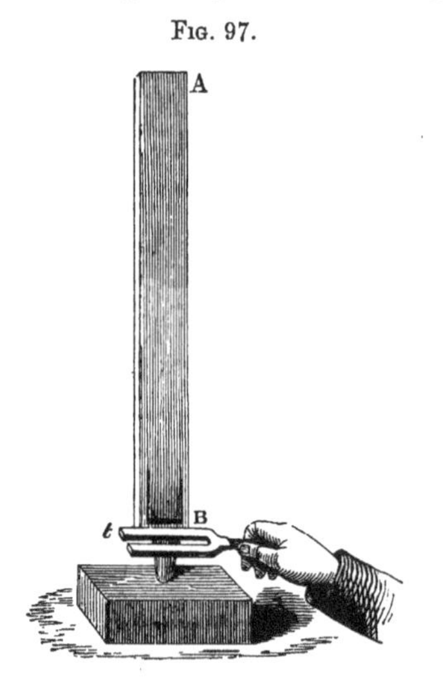

The heavy vibrating mass of the tuning-fork is practically uninfluenced by the motion of the air within the pipe. But this is not the case when air itself is the vibrating body. In this case, as before explained, the pipe creates, as it were, its own tuning-fork, by compelling the fluttering stream at its embouchure to vibrate in periods answering to its own.

We have now to enquire into the condition of the air within an open organ-pipe when its fundamental note is sounded. It is that of a rod free at both ends, held at its centre, and caused to vibrate longitudinally. The two ends are places of vibration, the centre is a node. How do I know this? Is there any way of *feeling* the vibrating column so as to determine its nodes and its places of vibration? The late excellent William Hopkins has taught us the following mode of solving this problem. I have here a little hoop, over which is stretched a thin membrane, capable of being readily thrown into vibration. The front of this organ-pipe, PP′, fig. 98, is of glass, through which you can see the position of any body within it. By means of a string, I can lower and raise at pleasure this little tambourine, *m*, through the entire length of the pipe. It now rests just above the upper end of the column of air. I sound the pipe. You hear immediately the loud buzzing of the membrane. I lower it into the pipe; it still continues to buzz; now its sound is becoming feeble; and now it has totally ceased. I cannot see the position of the membrane, but I venture to affirm that it is now in the middle of the pipe. It no longer vibrates,

FIG. 98.

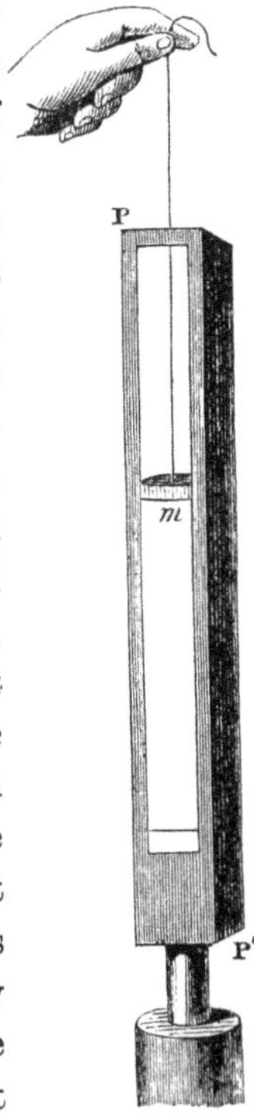

for the air around it is at rest. I lower it still further; the buzzing sound instantly commences, and continues down to the bottom of the pipe. Thus, as I lift the membrane, and lower it in quick succession, I have always, during every descent and ascent, two periods of sound separated from each other by one of silence. If sand were strewn upon the membrane it would dance above and below, but it would be quiescent at the centre. We thus prove experimentally, that when an organ-pipe sounds its fundamental note, it divides itself into two semi-ventral segments separated by a node.

What is the condition of the air at this node? Again, that of the middle of a rod, free at both ends, which yields the fundamental note of its longitudinal vibration. The pulses reflected from both ends of the rod, or of the column of air, meet in the middle, and produce compression; they then retreat and produce rarefaction. Thus, while there is no vibration in the centre, the air there undergoes the greatest changes of density. At the two ends of the pipe, on the other hand, the air-particles merely swing up and down without sensible compression or rarefaction.

If the sounding pipe were pierced at the centre, and the orifice stopped by a membrane, the air when condensed, would press the membrane outwards, and, when rarefied, the external air would press the membrane inwards. The membrane would therefore vibrate in unison with the column of air. The organ-pipe I now hold in my hand, fig. 99, is so arranged, that a small jet of gas, *b*, can be lighted opposite the centre of the pipe, and there acted upon by the vibrations of a membrane, such as I have described. There are two other gas-jets, *a* and *c*, placed midway between the centre and the two ends of the pipe. The three burners, *a*, *b*, *c*, are fed in the following manner:—through the tube, *t*, the gas enters the hollow chamber, *e d*, from which issue three little bent tubes, each communicating with a cavity closed underneath by the membrane, which is in direct contact with the air of the organ-pipe. From the three cavities issue the three little burners, *a*, *b*, *c*. I light them all; and now I blow into the pipe so as to sound its fundamental note. The three flames are agitated, but the central one is most so. I lower the flames and render them very small. If I now blow into the pipe, the central flame, *b*, will be extinguished, while the others will remain lighted. I ignite the gas-jets, and sound the fundamental note half-a-dozen times in succession; the middle flame is, in all cases, blown out.

Fig. 99.

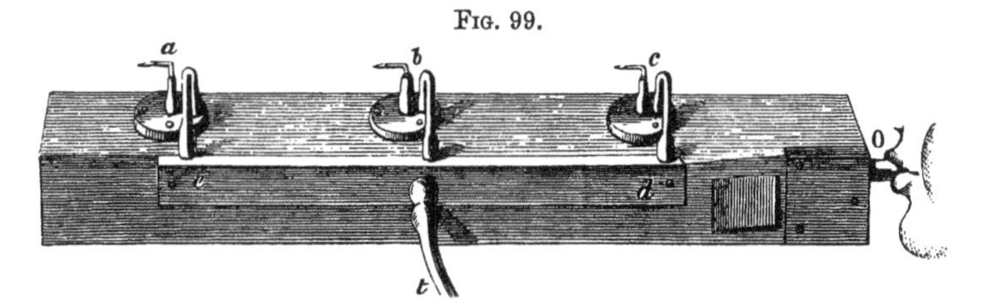

If I blow more sharply into this pipe, I cause the column of air to subdivide itself so as to yield its first overtone. The middle node no longer exists. The centre of the pipe is now a place of maximum vibration, while two nodes are formed midway between the centre and the two ends. But if this be the case, and if the flame opposite a node be always blown out, then, when I sound the first overtone of this pipe, the two flames *a* and *c* will be extinguished, while the central flame will remain lighted. I repeat the experiment three or four times, so as to assure you that there is no accident in the matter. When I sound the first harmonic, the two nodal flames are infallibly extinguished, while the flame *b* in the middle of the ventral segment is not sensibly disturbed.

There is no theoretic limit to the subdivision of an organ-pipe either stopped or open. In stopped pipes we begin with a semi-ventral segment, and pass on to 3,

5, 7, &c., semi-ventral segments; the number of vibrations of the successive notes augmenting in the same ratio. In open pipes we begin with 2 semi-ventral segments, and pass on to 4, 6, 8, 10, &c.; the number of vibrations of the successive notes augmenting in the same ratio; that is, in the ratio 1 : 2 : 3 : 4 : 5, &c. When, therefore, we pass from the fundamental tone to the first overtone in an open pipe, we obtain the octave of the fundamental. When we make the same passage in a stopped pipe, we obtain a note a fifth above the octave. No intermediate modes of vibration are in either case possible. If the fundamental tone of a stopped pipe be produced by 100 vibrations a second, the first overtone will be produced by 300 vibrations, the second by 500, and so on. Such a pipe, for example, cannot execute 200 or 400 vibrations in a second. In like manner the open pipe, whose fundamental note is produced by 100 vibrations a second, cannot vibrate 150 times in a second, but passes at a jump to 200, 300, 400, and so on.

In open pipes, as in stopped ones, the number of vibrations executed in the unit of time is inversely proportional to the length of the pipe. This follows from the fact, already dwelt upon so often, that the time of a vibration is determined by the distance which the sonorous pulse has to travel to complete a vibration.

In fig. 100, *a* and *b* represent the division of an open pipe corresponding to its fundamental tone; *c* and *d* represent the division corresponding to its first; *e* and *f* the division corresponding to its second overtone. The distance *m n*, is one-half, *o p* is one-fourth, and *s t* is one-sixth of the whole length of the pipe. These three lengths are therefore in the ratio of the three numbers 1 : 2 : 3. But the pitch of *a* is that of a stopped pipe equal in length to *m n*; the pitch of *c* is that of a stopped pipe of the length *o p*; while the pitch of *e* is that of a stopped pipe of the length *s t*. Hence, as these lengths are in the ratio of 1 : 2 : 3, the rates of vibration must be in the same proportion. From the mere inspection, therefore, of the respective modes of vibration we can draw the inference that the succession of tones in an open pipe must correspond to the series of natural numbers.

Fig. 100.

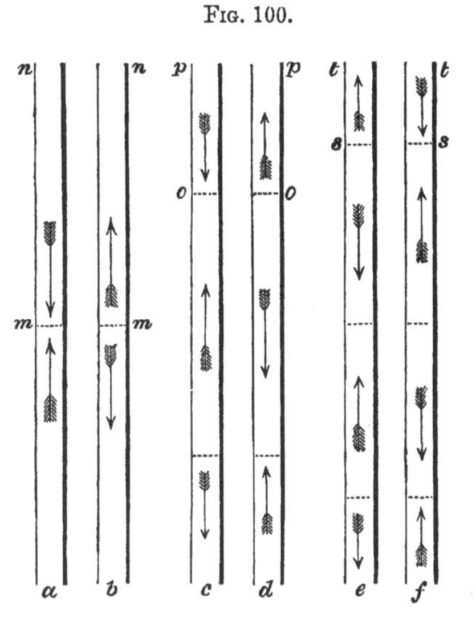

I have purposely drawn the pipe *a*, fig. 100, twice the length of *a*, fig. 93 (p. 180). It is perfectly manifest that to complete a vibration the pulse has to pass over the same distance in both pipes, and hence that the pitch of the two pipes must be the same. The open pipe, *a n*, consists virtually of two stopped ones, with the central nodal surface at *m* as their common base. The mere inspection, therefore, of the respective modes of vibration proves the relation of a stopped pipe to an open one to be that which experiment establishes.

If I have made these points intelligible you are now in a condition to grapple with problems which must at first sight appear absolutely insurmountable. We have already learned that the *relative* velocities of sound in different solid bodies may be determined from the notes which they emit when thrown into longitudinal vibration; and it was remarked at the time, that all we required to draw up a table of *absolute* velocities was the accurate comparison of the velocity in any one of those solids, with the velocity in air. We are now in a condition to supply what was then wanting. For we have learned that the vibrations of the air in an organ-pipe open at both ends are executed precisely as those of a rod free at both ends. Any difference of rapidity, therefore, between the vibrations of a rod and of an open organ-pipe, of the same length, must be due solely to the different velocities with which the sonorous pulses are propagated through them. Take therefore an organ-pipe of a certain length, emitting a note of a certain pitch, and find the length of a rod of pine which yields the same note. This length would be ten times that of the organ-pipe, which would prove the velocity of sound in pine to be ten times its velocity in air. But the absolute velocity in air is 1,090 feet a second; hence the absolute velocity in pine is 10,900 feet a second; which is that given in our first lecture (see p. 41). To the eminent Chladni we are indebted for this mode of determining the velocity of sound in solid bodies.

We had also in our first lecture a table of the velocities of sound in other gases than air. How was this table constructed? I am persuaded that you could answer me after due reflection. It would only be necessary to find a series of organ-pipes which, when filled with the different gases, yield the same note; the lengths of these pipes would give the relative velocities of sound through the gases. Thus we should find the length of a pipe filled with hydrogen to be four times that of a pipe filled with oxygen, and yielding the same note, which would prove the velocity of sound in the former to be four times its velocity in the latter.

But we had also a table of velocities through various liquids. How could it be constructed? By forcing the liquids through properly constructed organ-pipes, and comparing their musical tones. Thus, in water it would require a pipe a little better than four feet long to produce the note of a pipe in air one foot long, which proves that the velocity of sound in water is somewhat more than four times its velocity in air. My object here is to give you a clear notion of the way in which scientific knowledge enables us to cope with these apparently insurmountable problems. I do not go into the niceties of these measurements, because this is not needed for my present purpose. But you will readily comprehend that all the experiments with gases might be made with the same organ-pipe, the velocity of sound in each respective gas being immediately deduced from the pitch of its note. With a pipe of constant length the pitch, or, in other words, the number of vibrations, would be directly proportional to the velocity. Thus, comparing oxygen with hydrogen, we should find the note of the latter the double octave of that of the former, whence we should infer the velocity of sound in hydrogen to be four times its velocity in oxygen. The same remark applies to experiments with liquids. Here also the same pipe may be employed throughout, the velocities being inferred from the notes produced by the respective liquids.

In fact, the length of an open pipe being, as already explained, one-half the length of its sonorous wave, it is only necessary to determine, by means of the syren, the number of vibrations executed by the pipe in a second, and to multiply this number by twice the length of the pipe, in order to obtain the velocity of sound in the gas within the pipe. Thus, an open pipe 26 inches long and filled with air, executes 256 vibrations in a second. The length of its sonorous wave is $4\frac{1}{3}$ feet: multiply 256 by $4\frac{1}{3}$, we obtain 1,120 feet per second as the velocity of sound through air of this temperature. Were the tube filled with carbonic acid gas, its vibrations would be slower; were it filled with hydrogen, its vibrations would be quicker; and applying the same principle, we should find the velocity of sound in both these gases.

The application of this principle, moreover, is not restricted to gases. The length of an open pipe sounding in water, is half that of the sonorous wave which it generates in water. So likewise the length of a solid rod free at both ends, and sounding its fundamental note, is half that of the sonorous wave *in the substance of the solid.* Hence we have only to determine the rate of vibration of such a rod, and multiply it by twice the length of the rod, to obtain the velocity of sound in the substance of the rod. You can hardly fail to be impressed by the power which physical science has given us over these problems; nor will you refuse your admiration to that famous old investigator, Chladni, who taught us how to master them experimentally.

REEDS AND REED-PIPES.

The construction of the syren, and our experiments with that instrument, are, no doubt, fresh in your recollection. The musical sounds were there produced by the cutting up into puffs of a series of air currents. The same purpose is effected by a vibrating reed, as we find it employed in the accordion, the concertina, and the harmonica. In these instruments it is not the vibrations of the reed itself, which, imparted to the air, and transmitted through it to our organs of hearing, produce the music; the function of the reed is constructive, not generative; it moulds into a series of discontinuous puffs, that which without it would be a continuous current of air.

Reeds, if associated with organ-pipes, sometimes command, and are sometimes commanded by, the vibrations of the column of air. When they are stiff they rule the column; when they are flexible the column rules them. In the former case, to derive any advantage from the air-column, its length ought to be so regulated that either its fundamental tone or one of its overtones shall correspond

FIG. 101.

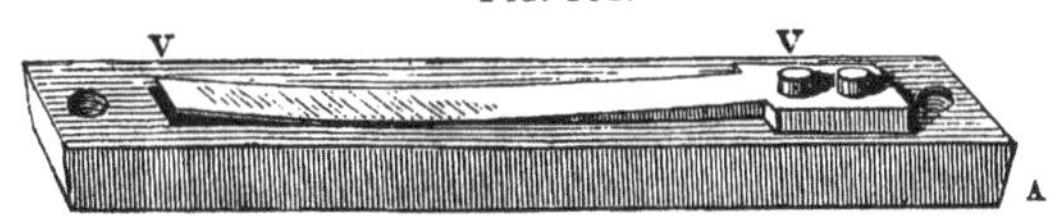

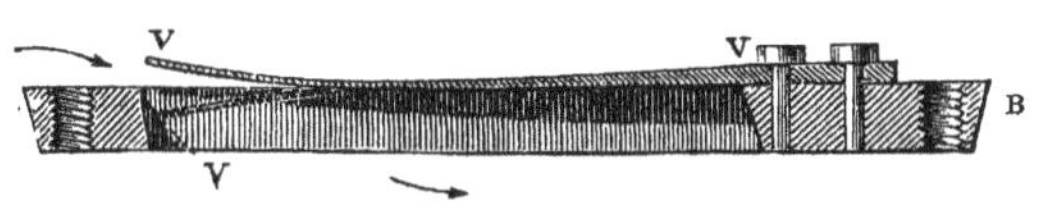

to the rate of vibration of the reed. The metal reed commonly employed in organ-pipes is shown in fig. 101, A and B, both in perspective and in section. It consists of a long and flexible strip of metal V, V, placed in a rectangular orifice through which the current of air enters the pipe. The manner in which the reed and pipe are associated is here shown. The front, *b c*, fig. 102, of the space containing the flexible tongue, is of glass, so that you may see the tongue within it. A conical pipe, A B. surmounts the reed.* The wire shown pressing against the reed is employed to lengthen or shorten it, and thus to vary within certain limits its rate of vibration. The reed formerly employed closed the aperture by simply falling against its sides; every stroke of the reed produced a tap, and these linked themselves together to an unpleasant, screaming

FIG. 102.

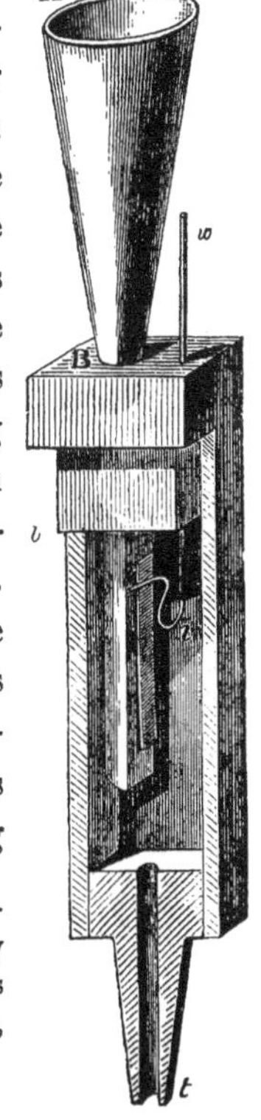

* The clear cuts of organ-pipes and reeds introduced here, and at p. 183, have been substantially copied from the excellent work of Helmholtz. Pipes opening with hinges so as to show their inner parts, were shown in the lecture.

sound, which materially injured that of the associated organ-pipe. This was mitigated, but not removed, by permitting the reed to strike against soft leather; but the reed now employed is the *free reed*, which vibrates to and fro between the sides of the aperture, almost, but not quite, filling it. In this way the unpleasantness referred to is avoided. When reed and pipe synchronise perfectly, the sound is most pure and forcible; a certain latitude, however, is possible on both sides of perfect synchronism. But if the discordance be pushed too far, the tube ceases to be of any use. We then obtain the sound due to the vibrations of the reed alone.

Flexible wooden reeds are also employed in organ-pipes, and such reeds can accommodate themselves to the requirements of the pipes above them. Perhaps the simplest illustration of the action of a reed which is commanded by its aërial column is furnished by a common wheaten straw. At about an inch from a knot I bury my penknife in this straw, *s r′*, fig. 103, to a depth of about one-fourth of the straw's diameter, and, turning the blade flat, I pass it upwards towards the knot, thus raising a strip of the straw nearly an inch in length. This strip,

Fig. 103.

r r′, is to be our reed, and the straw itself is to be our pipe. It is now eight inches long. I blow into it, and you hear this decidedly musical sound. I now cut it so as to make its length six inches; the pitch is higher: I make it four inches; the pitch is higher still. I make it two inches; the sound is now very shrill indeed. In all these experiments we had the same reed, which was compelled to accommodate itself throughout to the requirements of the vibrating column of air.

The clarionet is a reed pipe. It has a single broad tongue, with which a long cylindrical tube is associated. The reed end of the instrument is grasped by the lips, and by their pressure the slit between the reed and its frame is narrowed to the required extent. The overtones of a clarionet are different from those of a flute. A flute is an open pipe, a clarionet a stopped one, the end at which the reed is placed answering to the closed end of the pipe. The tones of a flute follow the order of the natural numbers, 1, 2, 3, 4, &c., or of the even numbers, 2, 4, 6, 8, &c.; while the tones of a clarionet follow the order of the odd numbers, 1, 3, 5, 7, &c. The intermediate notes are supplied by opening the lateral orifices of the instrument. Mr. Wheatstone was the first to make known this difference between the flute and clarionet, and his results agree with the recent investigations of Helmholtz. In the hautboy and bassoon we have two reeds inclined to each other at a sharp angle, with a slit between them, through which the air is urged. The pipe of the hautboy is *conical*, and its overtones are those of an open pipe,—different, therefore, from those of a clarionet. The pulpy end of a straw of green corn may be split by squeezing it, so as to form a double reed of this kind, and such a straw yields a musical tone. In the horn, trumpet, and serpent, the performer's lips play the part of the reed. These instruments are formed of long conical tubes, and their overtones are those of an open organ pipe. The music of the older instruments of this class was limited to their overtones, the particular tone elicited depending on the force of the blast and the tension of the lips. It is now usual to fill the gaps between the successive overtones by means of keys, which enable the performer to vary the length of the vibrating column of air.

The most perfect of reed instruments is the organ of voice. The vocal organ in man is placed at the top of the trachea or windpipe, the head of which is adjusted for the attachment of certain elastic bands which almost close the aperture. When the air is forced from the lungs through the slit which separates these *vocal chords*, they are thrown into vibration; by varying their tension, the rate of vibration is varied, and the sound changed in pitch. The vibrations of the vocal chords are practically unaffected by the resonance of the mouth, though we shall afterwards learn that this resonance, by reinforcing one or the other of the tones of the vocal chords, influences in a striking manner the clang-tint of the voice. The sweetness and smoothness of the voice depend on the perfect closure of the slit of the glottis at regular intervals during the vibration. Were the apparatus of the voice that now addresses you examined, it would doubtless appear, either that the edges of the vocal chords are more or less serrated, or that they strike each other, or that they imperfectly close the slit during their vibration, the harshness of tone which you tolerate so patiently being thus accounted for.

The vocal chords may be illuminated and viewed in a mirror placed suitably at the back of the mouth. I once sought to project the larynx of M. Czermak, upon a screen in this room, but with only partial success. The organ may, however, be viewed directly in the laryngoscope; its motions, both in singing, speaking, and coughing, being strikingly visible. It is represented at rest in fig. 104. The roughness of the voice in colds is due, according to Helmholtz, to mucous flocculi, which get into the slit of the glottis, and which are seen by means of the laryngoscope.

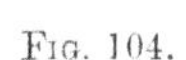

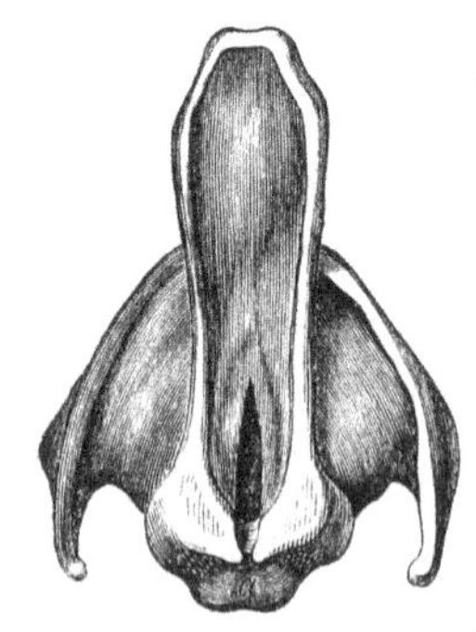

The squeaking falsetto voice with which some persons are afflicted, Helmholtz thinks may be produced by the drawing aside of the mucous layer which ordinarily lies under and loads the vocal chords. Their edges thus become sharper, and their weight less; while their elasticity remaining the same, they are shaken into more rapid tremors. The promptness and accuracy with which the vocal chords can change their tension, their form, and the width of the slit between them; to which must be added the elective resonance of the cavity of the mouth, renders the voice the most perfect of musical instruments.

The celebrated comparative anatomist, John Müller, imitated the action of the vocal chords by means of bands of india-rubber. He closed the open end of a glass tube by two strips of this substance, leaving a slit between them. On urging air through the slit, the bands were thrown into vibration, and a musical tone produced. Helmholtz recommends the form shown in fig. 105, where the tube, instead of ending in a section at right angles to its axis, terminates in two oblique sections, over which the bands of india-rubber are drawn. The easiest mode of obtaining sounds from reeds of this character is to roll round the end of a glass tube a strip of thin india-rubber, leaving about an inch of the substance projecting beyond the end of the tube. Taking two opposite portions of the projecting rubber in the fingers, and stretching it, a slit is formed, the blowing through which produces a musical sound, which varies in pitch, as the sides of the slit vary in tension.

Fig. 105.

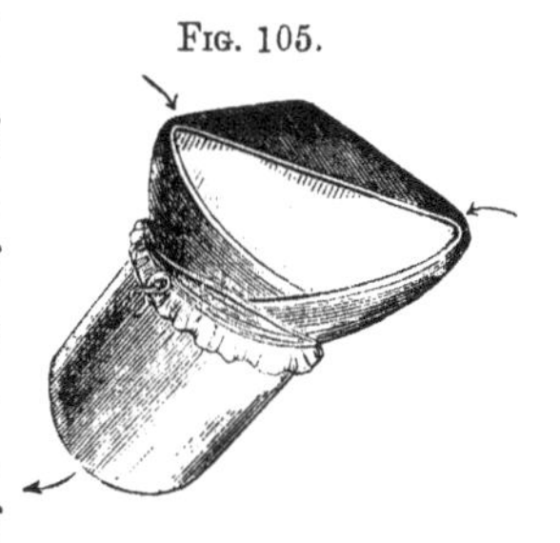

The formation of the vowel sounds of the human voice

excited long ago philosophic enquiry. We can distinguish one vowel sound from another, while assigning to both the same pitch and intensity. What, then, is the quality which renders the distinction possible? In the year 1779, this was made a prize question by the Academy of St. Petersburg, and Kratzenstein gained the prize for the successful manner in which he imitated the vowel sounds by mechanical arrangements. At the same time Von Kempelen, of Vienna, made similar and more elaborate experiments. The question was subsequently taken up by Mr. Willis, who succeeded, beyond all his predecessors, in the experimental treatment of the subject. The theory of vowel sounds was first stated by Mr. Wheatstone, and quite recently they have been made the subject of exhaustive enquiry by Helmholtz. With your present amount of knowledge you will find little difficulty in comprehending their origin.

Here, you observe, I have a free reed fixed in a frame, but without any pipe associated with it. I mount the reed on the acoustic bellows, and, urging air through its orifice, cause it to speak in this forcible manner. I now fix upon the frame of the reed a pyramidal pipe: you notice a change in the clang-tint, and by pushing my flat hand over the open end of the pipe, the similarity between the sounds produced and that of the human voice is unmistakable. Holding the palm of my hand over the end of the pipe, so as to close it altogether, and then raising my hand twice in quick succession, the word 'mamma' is heard as plainly as if it were uttered by an infant. For this pyramidal tube I now substitute a shorter one, and with it make the same experiment. The 'mamma' now heard is exactly such as would be uttered by a child with a stopped nose. Thus, by associating with a vibrating reed a suitable pipe, we can impart to the sound of the reed the qualities of the human voice.

Now, in the vocal organ of man, you have your reed in the vocal chords, and associated with this reed you have the resonant cavity of the mouth, which can so alter its shape as to resound, at will, either to the fundamental tone of the vocal chords or to any of their overtones. Through the agency of the mouth we can mix together the fundamental tone and the overtones of the voice in different proportions, and the different vowel sounds are due to different admixtures of this kind. I have here a series of tuning-forks, one of which I strike, and placing it before my mouth, adjust the size of that cavity until it resounds forcibly to the fork. When this is done, I remove the fork, and without altering in the least the shape or size of my mouth, I urge air through the glottis. I obtain the vowel sound U (*o o* in hoop) and no other. I take another fork, strike it, place it in front of the mouth, and adjust the cavity to resonance. After effecting this, I remove the fork and simply urge air through the glottis; I obtain the vowel sound O, and it is all that I can utter. Again, I take a third fork, adjust my mouth to it, and then urge air through the larynx; the vowel sound *ah!* and no other, is heard. In all these cases the vocal chords have been in the same constant condition. They have generated throughout the same fundamental tone and the same overtones, and the changes of sound which you have heard are due solely to the fact, that different tones in the different cases have been reinforced by the resonance of the mouth. Donders first proved that the mouth resounded differently for the different vowels.

Retaining the same fundamental tone, by the addition of other tones, or by the variation of the intensity of the fundamental, or of any particular member or members of its series of overtones, we can alter the quality of the clang, and it is by such additions and variations that we produce the different clang-tints of the human voice. It was enunciated by Ohm, and it has been experimentally proved by Helmholtz, that every clang, however complicated, may be reduced to a series of simple tones, beginning with the fundamental or deepest tone of the clang; and rising through others, whose rates of vibration are multiples of the rate of the fundamental. Let us call the fundamental tone of a clang the first tone, and let us call the tones whose rates of vibration correspond to twice, three times, four times, five times, &c., that of the fundamental, the second tone, the third tone, the fourth tone, the fifth tone, and so forth. This will enable us to define with greater clearness the manner in which such tones are blended in the production of vowel sounds.

For the production of the sound U (*o o* in hoop), I must push my lips forward, so as to make the cavity of the mouth as deep as possible, at the same time making the orifice of the mouth small. This arrangement corresponds to the deepest resonance of which the mouth is capable. The fundamental tone of the vocal chords is here reinforced, while the higher tones are thrown into the shade. The U is rendered a little more perfect when a feeble third tone is added to the fundamental.

The vowel O is pronounced when the mouth is so far opened that the fundamental tone is accompanied by its strong higher octave. A very feeble accompaniment of the third and fourth tones is advantageous, but not necessary.

The vowel A derives its character from the third tone, to strengthen which by resonance the orifice of the mouth must be wider, and the volume of air within it smaller than in the last instance. The second tone ought to be added in moderate strength, whilst weak fourth and fifth tones may also be included with advantage.

To produce E the fundamental tone must be weak, the second tone comparatively strong, the third very feeble, but the fourth, which is characteristic of this vowel, must be intense. A moderate fifth tone may be added. No essential change, however, occurs in the character of the sound when the third and fifth tones are omitted. In order to exalt the higher tones which characterise the vowel sound E, the resonant cavity of the mouth must be small.

In the production of the sound *Ah!* the higher overtones come principally into play; the second tone may be entirely neglected; the third rendered feebly; the higher tones, particularly the fifth and seventh, being added strongly.

These examples sufficiently illustrate the subject of vowel sounds. We may blend in various ways the elementary tints of the solar spectrum, producing innumerable composite colours by their admixture. Out of violet and red we produce purple, and out of yellow and blue we produce white. Thus also may elementary sounds be blended so as to produce all possible varieties of clang-tint. After having resolved the human voice into its constituent tones, Helmholtz was able to imitate these tones by tuning-forks, and, by combining them appropriately together, to produce the clang-tints of all the vowels.

Mistakes as to the fundamental tone of a clang may readily arise from the admixture of the higher tones. Before you is a heavy tuning-fork mounted on a large resonant case, and yielding, when excited, a deep and powerful tone. I stand near the case while the fork is sounding, and utter the vowel U (*o o* in hoop), in unison with the fork. The beats heard when I rise a little above, or fall a little below, the tone of the fork, declare the proximity of unison. I change the vowel sound from U to O; the beats are heard as before, which proves that the fundamental tone is unchanged in pitch: but you would be inclined to regard the sound of my voice as higher than before. The increase of sharpness is due, not to the heightening of the fundamental, but to the addition of the higher octave, which

Helmholtz has shown to be a necessary constituent of the O. The same remark applies, with even greater force, to the other vowels. Their apparent rise in pitch is due to the addition of the sharper overtones to the same constant fundamental. In her highest notes an opera-singer finds the vowel U a difficulty. The vowel E, for example, may be rendered after the utterance of U has ceased to be possible.

Unwilling to interrupt the continuity of our reasonings and experiments on the sounds of organ pipes, and their relations to the vibrations of solid rods, I have reserved for the conclusion of this discourse some reflections and experiments which, in strictness, belong to an earlier portion of the lecture. You have already heard the tones, and made yourselves acquainted with the various modes of division of a glass tube, free at both ends, when thrown into longitudinal vibration. When it sounds its fundamental tone, you know that the two halves of such a tube lengthen and shorten in quick alternation. If the tube were stopped at its ends, the vibrating stopper would throw the air within the tube into a state of vibration; and if the velocity of sound in air were equal to its velocity in glass, the air of the tube would vibrate in synchronism with the tube itself. But the velocity of sound in air is far less than its velocity in glass, and hence, if the column of air is to synchronise with the vibrations of the tube which contains it, it can only do so by dividing itself into vibrating segments of a suitable length. In an investigation of great interest, recently published in Poggendorff's Annalen, M. Kundt of Berlin has taught us how these segments may be rendered visible. I take a quantity of the light powder of lycopodium, and introducing it into this six-foot tube, shake it all over the interior surface: a small quantity of the powder clings to that surface. I stop the ends of the tube, clasp the middle in my hand, or, better still, I hold it at the centre by a fixed clamp, and sweep a wet cloth briskly over one of its halves. Instantly the powder, which a moment ago was clinging to its interior surface, falls to the bottom of the tube in the forms shown in fig. 106, the arrangement of the lycopodium marking the manner in which the column of air has

FIG. 106.

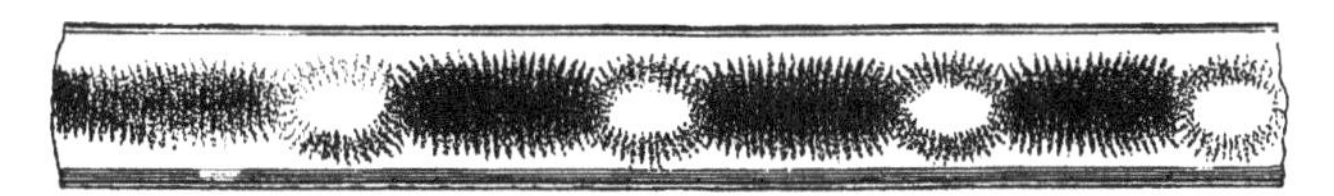

divided. Every node here is encircled by a ring of dust, while from node to node the dust arranges itself in transverse streaks along the ventral segments.

You will have little difficulty in seeing that we perform here, with air, substantially the same experiment as that of M. Melde with a vibrating string. When the string was too long to vibrate as a whole, it met the requirements of the tuning-fork to which it was attached by dividing into ventral segments. Now, in all cases, the length from a node to its next neighbour is half that of the sonorous wave: how many such half waves have we in our tube in the present instance? Sixteen (the figure shows only four of them). But the length of our glass tube vibrating thus longitudinally, is also half that of the sonorous wave *in glass*. Hence, in the case before us, with the same rate of vibration, the length of the semi-wave in glass is sixteen times the length of the semi-wave in air. In other words, the velocity of sound in glass is sixteen times its velocity in air. Thus, by a single sweep of the wet rubber, we solve a most important problem. But, as M. Kundt has shown, we need not confine ourselves to air. Introducing any other gas into the tube, a single stroke of our wet cloth enables us to determine the relative velocity of sound in that gas and in glass. When hydrogen is introduced, the number of ventral segments is less; when carbonic acid is introduced, the number is greater than in air.

From the known velocity of sound in air, coupled with the length of one of these dust segments, we can immediately deduce the number of vibrations executed in a second by the tube which contains the divided column of air. Clasping a glass tube at its centre and drawing my wetted cloth over one of its halves, I elicit this shrill note. The length of every dust segment, now within the tube, is 3 inches. Hence the length of the aërial sonorous wave corresponding to this note is 6 inches. But the velocity of sound in air of our present temperature is 1,120 feet per second; and this distance would embrace 2,240 of our sonorous waves. This number, therefore, expresses the rate of vibration per second, corresponding to the note just sounded.

Instead of damping the centre of the tube, and making it a nodal point, we may employ any other of its subdivisions. Laying hold of it, for example, at a point midway between its centre and one of its ends, and rubbing it properly, we know that it divides into three vibrating parts, separated by two nodes. We also know that in this division the note elicited is the octave of that heard when a single node is formed at the middle of the tube; for the vibrations are then twice as rapid. If, therefore, we divide the tube, having air within it, by two nodes instead of one, the number of ventral segments revealed by the lycopodium dust will be thirty-two instead of sixteen. The same remark applies, of course, to all other gases.

Filling a series of four tubes with air, carbonic acid, coal gas, and hydrogen, and then rubbing each so as to produce two nodes, M. Kundt found the number of dust segments formed within the tube in the respective cases to be as follows:—

Air	32 dust segments.
Carbonic acid . .	40 ,,
Coal gas . . .	20 ,,
Hydrogen . . .	9 ,,

The velocities of sound through the gases are inversely proportional to these numbers. Calling the velocity in air unity, the following fractions express the ratio of this velocity to those in the other gases:—

Carbonic acid . .	$\frac{32}{40} = 0{\cdot}8$
Coal gas . . .	$\frac{32}{20} = 1{\cdot}6$
Hydrogen . . .	$\frac{32}{9} = 3{\cdot}56$

Referring to a table introduced in our first lecture, we learn that Dulong, by a totally different mode of experiment, found the velocity in carbonic acid to be 0·97, and in hydrogen 3·8 times the velocity in air. The results of Dulong were deduced from the sound of organ pipes filled with the various gases; but here, by a process of the utmost simplicity, we arrive at a close approximation to his results. Dusting the interior surfaces of our tubes, filling them with the proper gases, and sealing their ends, they may be preserved for an indefinite length of time. By properly shaking one of them at any moment, its inner surface becomes thinly coated with the dust; and afterwards a single stroke of the wet cloth produces the division from which the velocity of sound in the gas may be immediately inferred.

Savart found that a spiral nodal line is formed round a tube or rod when it vibrates longitudinally, and Seebeck showed that this line was produced, not by longitudinal, but by secondary transversal vibrations. Now this spiral nodal line tends to complicate the division of the dust

Fig. 107.

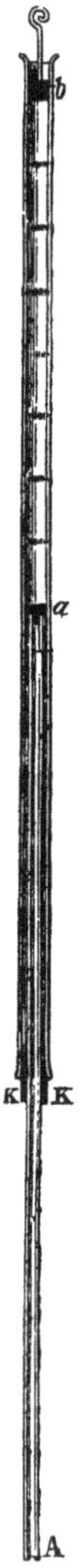

in our present experiments. It is, therefore, desirable to operate in a manner which shall altogether avoid the formation of this line; M. Kundt has accomplished this, by exciting the longitudinal vibrations in one tube, and producing the division into ventral segments in another which fits into the first like a piston. Before you is a tube of glass, fig. 107, seven feet long, and two inches internal diameter. One end of this tube is filled by the movable stopper, *b*. The other end, K K, is also stopped by a cork, through the centre of which passes the narrower tube, A *a*, which is firmly clasped at its middle by the cork, K K. The end of the interior tube, A *a*, is also closed with a projecting stopper, *a*, almost sufficient to fill the larger tube, but still fitting into it so loosely that the friction of *a* against the interior surface is too slight to interfere practically with the vibrations of the stopper. The interior surface between *a* and *b* being lightly coated with the lycopodium dust, I pass a wet cloth briskly over A K; instantly the dust between *a* and *b* divides into a number of ventral segments. When the length of the column of air, *a b*, is equal to that of the tube of glass, A *a*, the number of ventral segments is sixteen. When, as in the figure, *a b* is only one-half the length of A *a*, the number of ventral segments is eight.

But here you must perceive that the method of experiment is capable of great extension. Instead of the glass tube, A *a*, we may employ a rod of any other solid substance—of wood or metal, for example, and thus determine the relative velocity of sound in the solid and in air. In the place of the glass tube, I put a rod of brass of equal length, and rubbing its external half by a resined cloth, cause it to divide the column a b into the number of ventral segments proper to the metal's rate of vibration. In this way M. Kundt operated with brass, steel, glass, and copper, and his results prove the method to be capable of great accuracy. Calling, as before, the velocity of sound in air unity, in three different series of experiments, the following numbers expressive of the ratio of the velocity of sound in brass to its velocity in air were obtained:—

1st experiment	10·87
2nd experiment	10·87
3rd experiment	10·86

The coincidence is here extraordinary. To illustrate the possible accuracy of the method, I may add that the length of the individual dust segments was measured. In a series of twenty-seven experiments, this length was found to vary between 43 and 44 millimètres (each millimètre $\frac{1}{25}$th of an inch), never rising so high as the latter, and never falling so low as the former. The length of the metal rod, compared with that of one of the segments capable of this accurate measurement, gives us at once the velocity of sound in the metal, as compared with its velocity in air.

Three distinct experiments, performed in the same manner, gave the following velocities through steel, the velocity through air, as before, being regarded as unity:—

1st experiment	15·34
2nd experiment	15·33
3rd experiment	15·34

Here the coincidence is quite as perfect as in the case of brass.

In glass, by this new mode of experiment, the velocity was found to be

15·25.*

* The velocity in glass varies with the quality of the glass; the result of each experiment has therefore reference only to the particular kind of glass employed in the experiment.

Finally, in copper the velocity was found to be

11·96.

These results agree extremely well with those obtained by other methods. Wertheim, for example, found the velocity of sound in steel wire to be 15·108; M. Kundt finds it to be 15·34: Wertheim also found the velocity in copper to be 11·17; M. Kundt finds it to be 11·96. The differences are not greater than might be produced by differences in the materials employed by the two experimenters.

The length of the aërial column may or may not be an exact multiple of the wave length, corresponding to the rod's rate of vibration. If not, the dust segments take the form shown in fig. 108. But if, by means of the

FIG. 108.

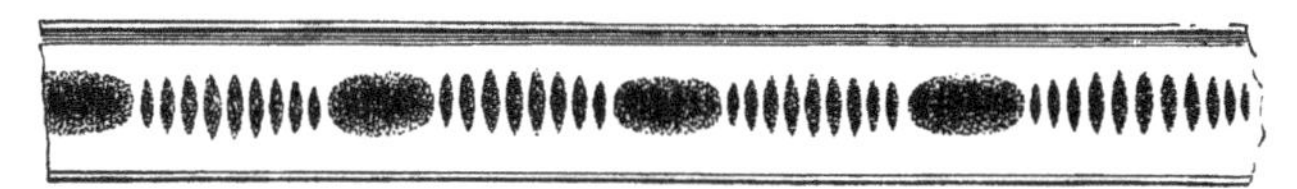

stopper, the column of air be made an exact multiple of the wave-length, then the dust quits the vibrating segments altogether, and forms, as in fig. 109, little isolated heaps at the nodes. And here a difficulty presents

FIG. 109.

itself. The stopped end *b* of the tube fig. 107 is, of course, a place of no vibration, where in all cases a nodal dust-heap is formed; but whenever the column of air was an exact multiple of the wave-length, M. Kundt always found a dust-heap close to the end *a* of the rod. Thus the point from which all the vibration emanated seemed itself to be a place of no vibration.

This difficulty was pointed out by M. Kundt, but he did

not attempt its solution. I trust we are now in a condition to explain it. In Lecture III., I had occasion to remark that in strictness a node is not a place of no vibration; that it is a place of minimum vibration; and that by the addition of the minute pulses which the node permits, vibrations of vast amplitude may be produced. The ends of M. Kundt's tube are such points of minimum motion, the lengths of the vibrating segments being such, that by the coalescence of direct and reflected pulses, the air at a distance of half a ventral segment from the end of the tube vibrates much more vigorously than that at the end of the tube itself. This addition of impulses is more complete when the aërial column is an exact multiple of the wave-length, and hence the reason why, in this case, the vibrations become intense enough to sweep the dust altogether away from the vibrating segments. The same point is illustrated by M. Melde's experiments, in which the tuning-forks, which are the sources of all the motion, are themselves nodes.

An experiment of Helmholtz's is here capable of instructive application. I hold in my hand a tuning-fork which executes 512 complete vibrations in a second. I stretch the string of the sonometer described in our third lecture over its two bridges, and place the iron stem of the tuning-fork upon the string. At present you hear no augmentation of the sound of the fork; the string remains quiescent. But I move the fork along the string, and here, at the number 33, a loud note suddenly issues from the string. With this particular tension the length 33 exactly synchronises with the vibrations of the fork. By the intermediation of the string, therefore, the fork is enabled to transfer its motion to the sonometer, and through it to the air. Maintaining the fork at the proper point, the sound continues as long as the fork vibrates. The least movement right or left from this point causes a sudden fall of the sound. I now tighten the string by means of the key. The note disappears; in fact it requires a greater length of this more highly tensioned string to respond to the fork. I therefore move the fork further away, and at the number 36 the note again bursts forth; tightening still more, I find 40 to be the point where the note is produced with maximum power. I now slacken the string. This slacker wire must be shortened in order to make it respond to the fork. I therefore move the fork towards the end of the string, and at the number 25 I find the note as before. Again I shift the fork to 35; nothing is heard; but by turning the key cautiously, I move the point of synchronism, if I may use the term, further from the end of the string. It finally reaches the fork, and at that moment a clear full note issues from the sonometer. In all cases, before the exact point is attained, and immediately in its vicinity, we hear 'beats,' which, as we shall afterwards understand, are due to the coalescence of the sound of the fork with that of the string, when they are nearly but not quite in unison with each other.

In all these experiments, when the length of wire between the tuning-fork and the bridge is such as to synchronise with the vibration of the fork, that vibration is transmitted through the wire to the box of the sonometer, and thence to the surrounding air. *In every case the point on which the fork rests is a nodal point.* It constitutes the comparatively fixed extremity of the wire whose vibrations synchronise with those of the fork. The case is exactly analogous to that of the hand holding the india-rubber tube, and to the tuning-fork in the experiments of M. Melde. It is also an effect precisely the same in kind as that observed by M. Kundt, where the end of the column of air in contact with his vibrating rod proved to be a node instead of the middle of a ventral segment.

ADDENDUM REGARDING RESONANCE.

The resonance of caves and of rocky enclosures is well known. Bunsen notices the thunder-like sound produced when one of the steam jets of Iceland breaks out near the mouth of a cavern. Most travellers in Switzerland have noticed the deafening sound produced by the fall of the Reuss at the Devil's Bridge. The noise of the fall is raised by resonance to the intensity of thunder. The sound heard when a hollow shell is placed close to the ear is a case of resonance. Children think they hear in it the sound of the sea. The noise is really due to the reinforcement of the feeble sounds with which even the stillest air is pervaded. By using tubes of different lengths, the variation of the resonance with the length of the tube may be noticed. The channel of the ear itself is also a resonant cavity. When a poker is held by two strings, and when the fingers of the hands holding the poker are thrust into the ears, on striking the poker against a piece of wood, a sound is heard as deep and sonorous as that of a cathedral bell. When open, the channel of the ear resounds to notes whose periods of vibration are about 3,000 per second. This has been shown by Helmholtz, and a German lady, named Seiler, has found that dogs which howl to music are particularly sensitive to the same notes.

SUMMARY OF LECTURE V.

When a stretched wire is suitably rubbed, in the direction of its length, it is thrown into longitudinal vibration: the wire can either vibrate as a whole or divide itself into vibrating segments separated from each other by nodes.

The tones of such a wire follow the order of the numbers 1, 2, 3, 4, &c.

The *transverse* vibrations of a rod fixed at both ends do not follow the same order as the transverse vibrations of a stretched wire; for here the forces brought into play, as explained in Lecture IV., are different. But the longitudinal vibrations of a stretched wire do follow the same order as the longitudinal vibrations of a rod fixed at both ends, for here the forces brought into play are the same, being in both cases the elasticity of the material.

A rod fixed at one end vibrates longitudinally as a whole, or it divides into two, three, four, &c. vibrating parts, separated from each other by nodes. The order of the tones of such a rod is that of the odd numbers 1, 3, 5, 7, &c.

A rod free at both ends can also vibrate longitudinally. Its lowest note corresponds to a division of the rod into two vibrating parts by a node at its centre. The overtones of such a rod correspond to its division into three, four, five, &c. vibrating parts, separated from each other by two, three, four, &c. nodes. The order of the tones of such a rod is that of the numbers 1, 2, 3, 4, 5, &c.

We may also express the order by saying that while the tones of a rod fixed at both ends follow the order of the odd numbers 1, 3, 5, 7, &c., the tones of a rod free at both ends follow the order of the even numbers 2, 4, 6, 8, &c.

At the points of maximum vibration the rod suffers no change of density; at the nodes, on the contrary, the changes of density reach a maximum. This may be proved by the action of the rod upon polarised light.

Columns of air of definite length resound to tuning-forks of definite rates of vibration.

The length of a column of air which most perfectly resounds to a fork is one-fourth of the length of the sonorous wave produced by the fork.

This resonance is due to the synchronism which exists between the vibrating period of the fork and that of the column of air.

By blowing across the mouth of a tube closed at one end, we produce a flutter of the air, and some pulse of this flutter may be raised by the resonance of the tube to a musical sound.

The sound is the same as that obtained when a tuning-fork, whose rate of vibration is that of the tube, is placed over the mouth of the tube.

When a tube closed at one end—a stopped organ-pipe for example—sounds its lowest note, the column of air within it is undivided by a node. The overtones of such a column correspond to its division into parts, like those of a rod fixed at one end and vibrating longitudinally. The order of its tones is that of the odd numbers 1, 3, 5, 7, &c. That this must be the order follows from the manner in which the column is divided.

In organ-pipes the air is agitated by causing it to issue from a narrow slit, and to strike upon a cutting edge. Some pulse of the flutter thus produced is raised by the resonance of the pipe to a musical sound.

When, instead of the aërial flutter, a tuning-fork of the proper rate of vibration is placed at the embouchure of an organ-pipe, the pipe *speaks* in response to the fork. In practice, the organ-pipe virtually creates its own tuning-fork, by compelling the sheet of air at its embouchure to vibrate in periods synchronous with its own.

An open organ-pipe yields a note an octave higher than that of a closed pipe of the same length. This relation is a necessary consequence of the respective modes of vibration.

When, for example, a stopped organ-pipe sounds its deepest note, the column of air, as already explained, is undivided. When an open pipe sounds its deepest note, the column is divided by a node at its centre. The open pipe in this case virtually consists of two stopped pipes with a common base. Hence it is plain that the note of an open pipe must be the same as that of a stopped pipe of half its length.

The length of a stopped pipe is one-fourth that of the sonorous wave which it produces, while the length of an open pipe is one-half that of its sonorous wave.

The order of the tones of an open pipe is that of the even numbers 2, 4, 6, 8, &c., or of the natural numbers 1, 2, 3, 4, &c.

In both stopped and open pipes the number of vibrations executed in a given time is inversely proportional to the length of the pipe.

The places of maximum vibration in organ-pipes are places of minimum change of density; while at the places of minimum vibration the changes of density reach a maximum.

The velocities of sound in gases, liquids, and solids may be inferred from the tones which equal lengths of them produce; or they may be inferred from the lengths of these substances which yield equal tones.

Reeds, or vibrating tongues, are often associated with vibrating columns of air. They consist of flexible laminæ which vibrate to and fro in a rectangular orifice, thus rendering intermittent the air-current passing through the orifice.

The action of the reed is the same as that of the syren.

The flexible wooden reeds sometimes associated with organ-pipes are compelled to vibrate in unison with the column of air in the pipe; other reeds are too stiff to be thus controlled by the vibrating air. In this latter case the column of air is taken of such a length that its vibrations synchronise with those of the reed.

By associating suitable pipes with reeds we impart to their tones the qualities of the human voice.

The vocal organ in man is a reed instrument, the vibrating reed in this case being elastic bands placed at the top of the trachea, and capable of various degrees of tension.

The rate of vibration of these vocal chords is practically uninfluenced by the resonance of the mouth; but the mouth, by changing its shape, can be caused to resound to the fundamental tone, or to any of the overtones of the vocal chords.

By the strengthening of particular tones through the resonance of the mouth, the clang-tint of the voice is altered.

The different vowel sounds are produced by different admixtures of the fundamental tone and the overtones of the vocal chords.

When the solid substance of a tube stopped at one, or at both ends, is caused to vibrate longitudinally, the air within it is also thrown into vibration.

By covering the interior surface of the tube with a light powder the manner in which the aërial column divides itself may be rendered apparent. From the division of the column the velocity of sound in the substance of the tube, compared with its velocity in air, may be inferred.

Other gases may be employed instead of air, and the velocity of sound in these gases, compared with its velocity in the substance of the tube, may be determined.

The end of a rod vibrating longitudinally may be caused to agitate a column of air contained in a tube, compelling the air to divide itself into ventral segments. These segments may be rendered visible by light powders, and from them the velocity of sound in the substance of the vibrating rod, compared with its velocity in air, may be inferred.

In this way the relative velocities of sound in all solid substances capable of being formed into rods, and of vibrating longitudinally, may be determined.

LECTURE VI.

SOUNDING FLAMES—INFLUENCE OF THE TUBE SURROUNDING THE FLAME—INFLUENCE OF SIZE OF FLAME—HARMONIC NOTES OF FLAMES—EFFECT OF UNISONANT NOTES ON SINGING FLAMES—ACTION OF SOUND ON NAKED FLAMES—EXPERIMENTS WITH FISH-TAIL AND BAT'S-WING BURNERS—EXPERIMENTS ON TALL FLAMES—EXTRAORDINARY DELICACY OF FLAMES AS ACOUSTIC REAGENTS—THE VOWEL FLAME—ACTION OF CONVERSATIONAL TONES UPON FLAMES—ACTION OF MUSICAL SOUNDS ON UNIGNITED JETS OF GAS—CONSTITUTION OF WATER JETS—ACTION OF MUSICAL SOUNDS ON WATER JETS—A LIQUID VEIN MAY COMPETE IN POINT OF DELICACY WITH THE EAR.

FRICTION is always rhythmic. When we pass a resined bow across a string, the tension of the string secures the perfect rhythm of the friction. When we pass the wetted finger round the edge of a glass, the breaking up of the friction into rhythmic pulses expresses itself in music. Savart's experiments prove the friction of a liquid against the sides of an orifice through which it passes to be competent to produce musical sounds. We have here the means of repeating his experiment. The tube A B, fig. 110, is filled with water, its extremity, B, being closed by a plate of brass, which is pierced by a circular orifice of a diameter equal to the thickness of the plate. Removing a little peg which stops the orifice, the water issues from it, and as it sinks in the tube a musical note of great sweetness issues from the liquid column. This note is due to the intermittent flow of the liquid through the orifice, by which the whole column above it is thrown into vibration. The tendency to this effect shows itself when tea is poured from a teapot, in the circular ripples that cover the falling liquid. The same intermittence is observed in the black dense smoke which rolls in rhythmic rings from the funnel of a steamer. The unpleasant noise of unoiled machinery is also a declaration of the fact that the friction is not uniform, but is due to the alternate 'bite' and release of the rubbing surfaces.

FIG. 110.

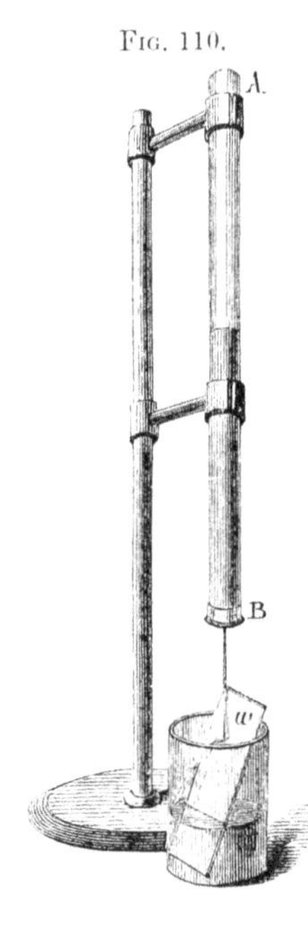

Where gases are concerned friction is of the same intermittent character. A rifle bullet sings in its passage through the air; while to the rubbing of the wind against the boles and branches of the trees are to be ascribed the 'waterfall tones' of an agitated pine-wood. Pass a steadily burning candle rapidly through the air; an indented band of light, declaring intermittence, is the consequence, while the almost musical sound which accompanies the appearance of this band is the audible expression of the rhythm. On the other hand, if you blow gently against a candle flame, the fluttering noise announces a rhythmic action. We have already learned what can be done when a pipe is associated with such a flutter; we have learned that the pipe selects a special pulse from the flutter, and raises it by resonance to a musical sound. In a similar manner the noise of a flame may be turned to account. The blowpipe flame of our laboratory, for example, when enclosed within an appropriate tube, has its flutter raised to a musical *roar*. The special pulse first selected soon reacts upon the flame so as to abolish in a great degree the other pulses, compelling the flame to vibrate in periods answering to the selected one. And this reaction can become so powerful—the timed shock of the reflected pulses may accumulate to such an extent—as to beat the flame, even when very large, into extinction.

Nor is it necessary to produce this flutter by any extraneous means. When a gas flame is simply enclosed within a tube, the passage of the air over it is usually sufficient to produce the necessary rhythmic action, so as to cause the flame to burst spontaneously into song. Not all, however, are aware of the intensity to which this flame-music may rise. I have here a ring burner with twenty-eight orifices, from which issues a gas flame. I place over the flame this tin tube, 5 feet long, and 2½ inches in diameter. The flame flutters at first, but it soon chastens its impulses into perfect periodicity, and a deep and clear musical note is the result. The quickness of its pulses depends in some measure on the size of the flame, and by lowering the gas I finally stop the note which is now sounded. After a momentary interval of silence, another note, which is the octave of the last, is yielded by the flame. The first note was the fundamental note of the tube which surrounds the flame: this is the first harmonic. In fact, here, exactly as in the case of open organ-pipes, we have the aërial column dividing itself into vibrating segments, separated from each other by nodes.

Permit me now to try the effect of this larger tube, *a b*, fig. 111, 15 feet long, and 4 inches wide, which was formed for a totally different use. It is supported by a steady stand *s s′*, and into it is lifted the tall burner, shown enlarged at B. You hear the incipient flutter; you now hear the more powerful sound. As the flame is lifted higher the action becomes more violent, until finally a storm of music issues from the tube. And now all has suddenly ceased;

Fig. 111.

the reaction of its own pulses upon the flame has beaten it into extinction. I now relight the flame and make it very small. When raised within the tube, the flame again sings, but it is one of the harmonics of the tube that you now hear. On turning the gas fully on, the note ceases —all is silent for a moment; but the storm is brewing, and soon it bursts forth, as at first, in a kind of hurricane of sound. By lowering the flame the fundamental note is abolished, and now you hear the first harmonic of the tube. Making the flame still smaller, the first harmonic disappears, and the second is heard. Your ears being disciplined to the apprehension of these sounds, I turn the gas once more fully on. Mingling with the deepest note you notice the harmonics, as if struggling to be heard amid the general uproar of the flame.

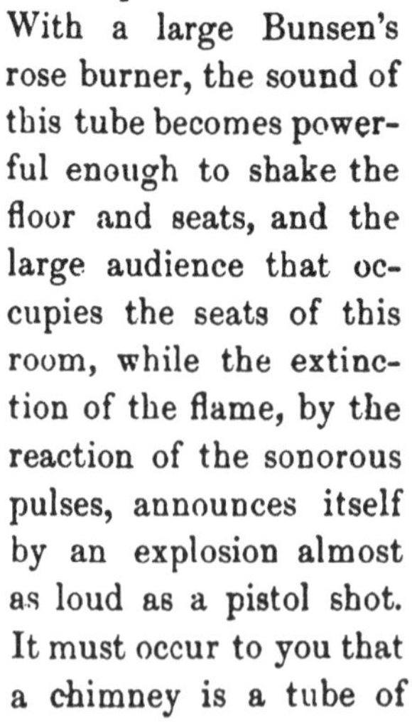

With a large Bunsen's rose burner, the sound of this tube becomes powerful enough to shake the floor and seats, and the large audience that occupies the seats of this room, while the extinction of the flame, by the reaction of the sonorous pulses, announces itself by an explosion almost as loud as a pistol shot. It must occur to you that a chimney is a tube of this kind upon a large scale, and that the roar of a flame in a chimney is simply a rough attempt at music.

I now pass on to shorter tubes and smaller flames. Here is a series of eight of them. Placing the tubes over the flames, each of them starts into song, and you notice that as the tubes lengthen the tones deepen. The lengths of these tubes are so chosen that they yield in succession the eight notes of the gamut. Round some of them you observe a paper slider, *s*, fig. 112, by which the resounding tube can be lengthened or shortened. While the flame is sounding I raise the slider; the pitch instantly falls. I now lower the slider; the pitch instantly rises. These experiments prove the flame to be governed by the tube. By the reaction of the pulses, reflected back upon the flame, its flutter is rendered perfectly periodic, the length of the period being determined, as in the case of organ-pipes, by the length of the tube.

Fig. 112.

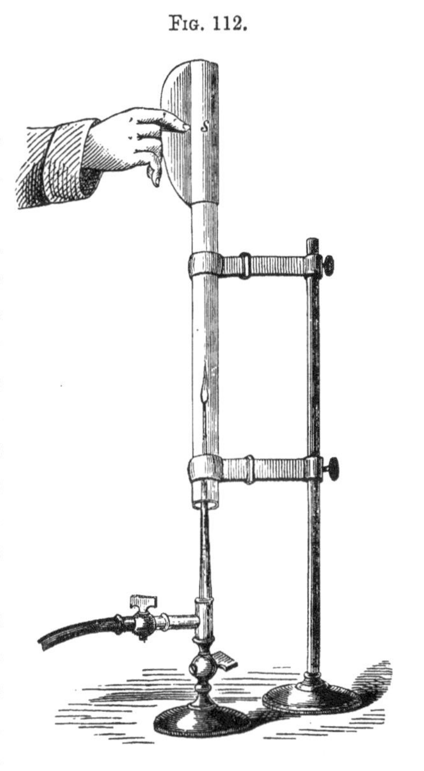

The fixed stars, especially those near the horizon, shine with an unsteady light, sometimes changing colour as they twinkle. I have often watched at night, upon the plateaux of the Alps, the alternate flash of ruby and emerald in the lower and larger stars. Place a piece of looking-glass so that you can see in it the image of such a star; on tilting the glass quickly to and fro, the line of light obtained will not be continuous, but will form a string of coloured beads of extreme beauty. You obtain the same effect when you point an opera-glass at the star and shake it. This experiment teaches that in the act of twinkling the light of the star is quenched at intervals: the dark spaces between the bright beads corresponding to the periods of extinction. Now our singing flame is a twinkling flame. You observe a certain quivering motion when it begins to sing, and you may analyse that motion with a looking-glass, or an opera-glass, as in the case of the star.* I can now see the image of this flame in a small looking-glass. On tilting the glass, so as to cause the image to form a circle of light, the luminous band is not continuous, as it would be if the flame were perfectly steady; it is resolved into a beautiful chain of flames, fig. 113.

With a larger, brighter, and less rapidly vibrating flame, you may all see this intermittent action. Over this gas flame, *f*, fig. 114, is placed a glass tube A B, 6 feet long, and 2 inches in diameter. The tube is in part blackened, so as to prevent the light of the flame from falling directly upon the screen, which it is now desirable to have as dark as possible. In front of the tube is placed a concave mirror, M, which forms upon the screen an enlarged image of the flame. I can turn the mirror with my hand and cause the image to pass over the screen. Were the flame silent and steady, we should obtain a *continuous* band of light;

FIG. 113.

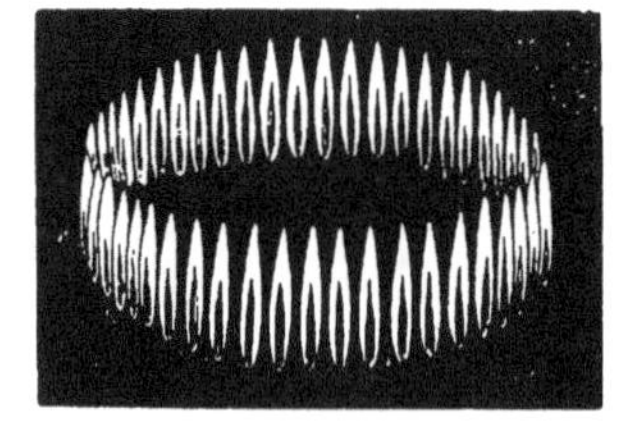

* This experiment was first made with a hydrogen flame by Mr. Wheatstone.

but it quivers, and emits at the same time this deep and powerful note. On twirling the mirror, we obtain, instead of a continuous band, a series of images, *o p*, forming a luminous chain. By turning fast, I separate these

FIG. 114.

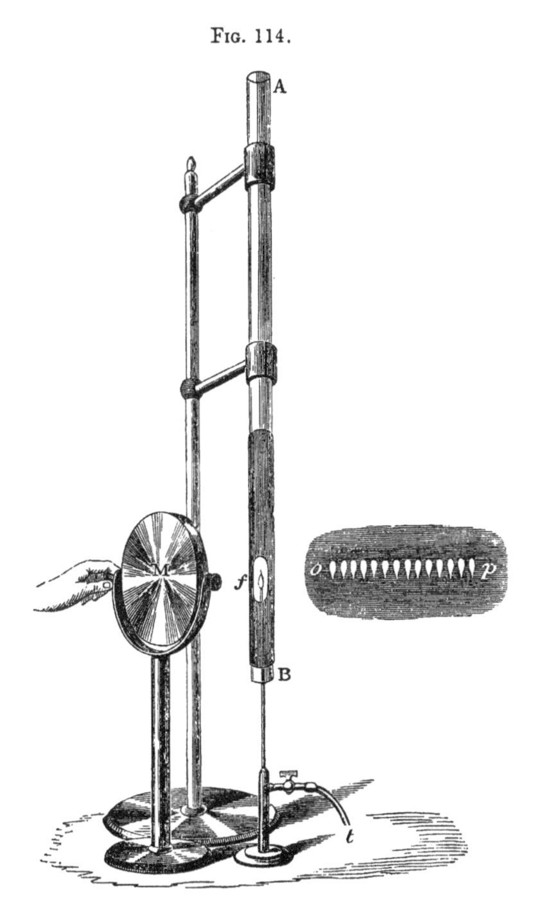

images more widely apart; by turning slowly I cause them to close up, and thus the chain of flames is caused to undergo the most beautiful variations. Clasping the lower end, B, of the tube with my hand, I so impede the air as to stop the flames' vibration. You have now a continuous band when the mirror is turned. Observe the suddenness with which this band breaks up into a rippling line of images the moment my hand is removed, and the current of air is permitted to pass over the flame.

When a *small* vibrating coal-gas flame is carefully examined by the rotating mirror, the beaded line is a series of flame-images, each consisting of yellow tip resting upon a base of the richest blue. In some cases I have been unable to observe any union of one flame with another; the spaces between the flames being absolutely dark to the eye. But if dark, the flame must have been totally extinguished at intervals, a residue of heat, however, remaining sufficient to reignite the gas. This I believe to be possible, for gas may be ignited by non-luminous air.* By means of the syren, we can readily determine the number of times this flame extinguishes and relights itself in a second. As the note of the instrument approaches that of the flame, unison is preceded by these well-known beats, which become gradually less rapid, and now the two notes melt into perfect unison, blending together to a single stream of song. I will endeavour to maintain the syren at this pitch for a minute. At the end of this time I find recorded upon our dials 1,700 revolutions. But the disc being perforated by 16 holes, it follows that every revolution corresponds to 16 pulses. Multiplying 1,700 by 16, we find the number of pulses in a minute to be 27,200. This number of times did our little flame extinguish and rekindle itself during the continuance of our experiment, that is to say, it was put out and relighted 453 times in a second.

A singing flame yields so freely to the pulses which

* A gas-jet, for example, can be ignited five inches above the tip of a visible gas-flame, where platinum leaf shows no redness.

fall upon it that it is almost wholly governed by the tube which surrounds it; *almost*, but not altogether. The pitch of the note depends in some measure upon the size of the flame. This is readily proved, by causing two flames to emit the same note, and then slightly altering the size of either of them. The unison is instantly disturbed by beats. By altering the size of a flame we can also, as already illustrated, draw forth the harmonic overtones of the tube which surrounds it. This experiment is best performed with hydrogen, its combustion being much more vigorous than that of ordinary gas. When this glass tube, which is nearly 7 feet long, is placed over a large hydrogen flame, the fundamental note of the tube is obtained, corresponding to a division of the column of air within it by a single node at the centre. Placing this second tube, 3 feet 6 inches long, over the same flame, no musical sound whatever is obtained; this large flame, in fact, is not able to accommodate itself to the vibrating period of the shorter tube. But, on lowering the flame, it soon bursts into vigorous song, its note being the octave of that yielded by the longer tube. I now remove the shorter tube, and once more cover the flame with the longer one. The longer tube no longer sounds its fundamental note, but the precise note of the shorter one. To accommodate itself to the vibrating period of the diminished flame, the longer column of air divides itself in the manner of an open organ-pipe when it yields its first harmonic. By varying the size of the flame, it is possible, with the tube now before you, to obtain a series of notes whose rates of vibration are in the ratio of the numbers 1 : 2 : 3 : 4 : 5, that is to say, the fundamental tone and its first four harmonics.

These sounding flames, though probably never before raised to the intensity in which they have been exhibited here to-day, are of old standing. In 1777, the sounds of a hydrogen flame were heard by Dr. Higgins. In 1802, they were investigated to some extent by Chladni, who also refers to an incorrect account of them given by De Luc. Chladni showed that the tones are those of the open tube which surrounds the flame, and he succeeded in obtaining the two first harmonics. In 1802 G. De la Rive experimented on this subject. Placing a little water in the bulb of a thermometer, and heating it, he showed that musical tones of force and sweetness could be produced by the periodic condensation of the vapour in the stem of the thermometer. He accordingly referred the sounds of flames to the alternate expansion and condensation of the aqueous vapour produced by the combustion. We can readily imitate his experiments. Holding a thermometer bulb containing water in the flame of a spirit lamp, with its stem oblique, the sounds are heard, soon after the water begins to boil. In 1818, however, Mr. Faraday showed that the tones were produced when the tube surrounding the flame was placed in air of a temperature higher than 100° C. He also showed that the tones could be obtained from flames of carbonic oxide, where aqueous vapour was entirely out of the question.

After these experiments, the first novel acoustic observation on flames was made in Berlin by the late Count Schaffgotsch, who showed that when an ordinary gas flame was surmounted by a short tube, a strong falsetto voice pitched to the note of the tube, or to its higher octave, caused the flame to quiver. In cases where the note of the tube was sufficiently high, the flame could even be extinguished by the voice.

In the spring of 1857, this experiment came to my notice, and I endeavoured to repeat it. No directions were given in the short account of the observation published in Poggendorff's Annalen; but, in endeavouring to ascertain the conditions of success, a number of singular effects forced themselves upon my attention. Meanwhile, Count Schaffgotsch also followed up the subject. To a great extent we travelled over the same ground, neither of us knowing how the other was engaged; but so far as the experiments then executed are common to us both, to Count Schaffgotsch belongs the priority.

Let me here repeat his first observation. Within this tube, 11 inches long, burns tranquilly a small gas flame. It is bright and silent. I have ascertained by trial the note of this tube, and now, standing at some distance from the flame, I sound that note; the flame quivers in a manner visible to you all. To obtain the extinction of the flame it is necessary to employ a burner with a very narrow aperture, from which the gas issues under considerable pressure. The little flame now burning before you is thus obtained. On sounding the note of the tube surrounding the flame, it quivers as before. I throw more power into my voice, and now the flame is extinguished.

The cause of the quivering of the flame will be best revealed by an experiment with the syren. Mounting the instrument on the acoustic bellows at some distance from a singing flame, I gradually exalt the pitch of the syren. As the note approaches that of the flame you hear beats, and at the same time you observe a dancing of the flame synchronous with the beats. The jumps succeed each other more slowly as unison is approached. They cease when the unison is perfect, and they begin again as soon as the syren is urged beyond unison, becoming more rapid as the discordance is increased. The quiver of the flame observed by M. Schaffgotsch was due to this cause. The flame jumped because the note of the tube surrounding it was nearly, but not quite in unison with the experimenter's voice.

That the jumping of the flame proceeds in exact accordance with the beats is well shown by a tuning-fork, which yields the same note as the flame. Loading such a fork with a bit of wax, so as to throw it slightly out of unison; and bringing it, when agitated, near the tube in which the flame is singing, the beats and the leaps occur at the same intervals. When the fork is placed thus over a resonant jar, all of you can hear the beats, and see at the same time the dancing of the flame. By changing the load upon the tuning-fork, or by slightly altering the size of the flame, the rate at which the beats succeed each other may be altered; but in all cases the jumps address the eye at the moments when the beats address the ear.

While executing these experiments, I once noticed that, on raising my voice to the proper pitch, a flame which had been burning silently in its tube began to sing. The song was interrupted, and the proper note sounded several times in succession; in every case the flame responded by starting into song. The same observation had, without my knowledge, been made a short time previously by Count Schaffgotsch. Observe the conditions of the experiment. I place a tube, 12 inches long, over this flame, which occupies a position within the tube about an inch and a half from its lower end. When the proper note is sounded the flame trembles, but it does not sing. I lower the tube, so that the flame shall be three inches from its lower end; it bursts spontaneously into song. Now, between these two positions there is a third, at which, if the flame be placed, it will burn silently; but if it be excited by the voice it will sing, and continue to sing.

In this position, then, it is able to sing, but it requires a start. It is as it were on the brink of a precipice, but it requires to be pushed over. I place the flame in this position; it is silent: but on the sounding of the proper note, it stretches forth its little tongue and begins its song. By placing my finger for an instant on the end of the tube I stop the music; and now, standing as far from the flame as this room will allow me, I command the flame to sing. It obeys immediately. I turn my back towards it, and repeat the experiment. My body does not shade the flame. The sonorous pulses run round me, reach the tube, and call forth the song. A pitch-pipe, or any other instrument which yields a note of the proper height, produces the same effect.

I now place three flames in three tubes *a*, *b*, *c*, 10, 12, and 14 inches in length respectively. The flames are silent; let us cause the sound of the syren to act upon them. The tone of the instrument gradually rises; it now approaches the pitch of the longest tube *c*, in passing which it calls forth the song of the flame within. Urging it still higher, we soon attain the pitch of *b*, which then starts into song. It now passes *a*, and, like the other two, the flame within *a* responds. Mounting, in this way, a series of tubes capable of emitting all the notes of the gamut, over suitable flames, with an instrument sufficiently powerful, and from a distance of 20 or 30 yards, a musician, by running over the gamut, might call forth every note in succession, the whole series of flames finally joining in the song.

When a silent flame, capable of being excited in the manner here described, is looked at in a moving mirror, it produces there a continuous band of light. Nothing can be more beautiful than the sudden breaking up of this band into a string of richly luminous pearls at the instant the voice is pitched to the proper note.

I will conclude these experiments on singing flames by making one flame effect the musical ignition of another. Before you are two small flames, *a* and *b*, a yard asunder, the tube over *a* being 10½ inches, and that over *b* 12 inches long. The shorter tube is clasped by a paper slider, which enables me to vary the note of the tube. The flame *a* is now singing, but the flame *b*, in the longer tube, is silent. I raise the paper slider which surrounds *a*, so as to lengthen the tube. Flame calls to flame; but as yet there is no response. The moment, however, I approach near enough to the pitch of the tube surrounding *b*, that flame sings. The experiment may be varied by making *b* the singing flame, and *a* the silent one at starting. Drawing out the telescopic slider, a point is soon attained where the flame *a* commences its song. In this way one flame may address another through considerable distances; when the arrangement is sufficiently sensitive, the proper note always elicits a response. I may add that it is also possible to silence the singing flame by the proper management of the voice.

SENSITIVE NAKED FLAMES.

We have hitherto dealt with flames surrounded by resonant tubes; and none of these flames, if naked, would respond in any way to such noise or music as could be here applied. Still it is possible to make naked flames thus sympathetic. In a former lecture (p. 101), I referred to the oscillations of water in a bottle, as revealing the existence of vibrations of a definite period in the general jar of a railway train. The fish-tail flames in some of our metropolitan railway carriages are far more sensitive acoustic reagents. If you pay the requisite attention, you will find single flames here and there jumping in synchronism with certain tremors of the train. A flame, for example, having a horizontal edge, when the train is still, will, during the motion, periodically thrust forth a central tongue, and continue to jump as long as a special set of vibrations is present. It will subside when those vibrations disappear, and jump again when they are restored. When the train is at rest, the tapping of the glass shade which surrounds the flame rarely fails, when it is sensitive, to cause it to jump.

This action of sound upon a naked fish-tail flame

was first observed by Professor Leconte at a musical party in the United States. His observation is thus described:—'Soon after the music commenced, I observed that the flame exhibited pulsations which were *exactly synchronous* with the audible beats. This phenomenon was very striking to every one in the room, and especially so when the strong notes of the violoncello came in. It was exceedingly interesting to observe how perfectly even the *trills* of this instrument were reflected on the sheet of flame. *A deaf man might have seen the harmony.* As the evening advanced, and the diminished consumption of gas in the city *increased the pressure*, the phenomenon became more conspicuous. The *jumping* of the flame gradually increased, became somewhat irregular, and, finally, it began to flare continuously, emitting the characteristic sound indicating the escape of a greater amount of gas than could be properly consumed. I then ascertained by experiment, that the phenomenon *did not* take place unless the discharge of gas was so regulated that the flame approximated to the condition of *flaring*. I likewise determined by experiment, that the effects *were not* produced by jarring or shaking the floor and walls of the room by means of repeated concussions. Hence it is obvious that the pulsations of the flame *were not* owing to *indirect* vibrations propagated through the medium of the walls of the room to the burning apparatus, but must have been produced by the *direct* influence of aërial sonorous pulses on the burning jet.'*

The significant remark, that the jumping of the flame was not observed until it was near flaring, suggests the means of repeating the experiments of Dr. Leconte; while a more intimate knowledge of the conditions of success enables us to vary and exalt them in a striking

* Philosophical Magazine, March 1858, p. 235.

degree. Before you burns a bright candle-flame: I may shout, clap my hands, sound this whistle, strike this anvil with a hammer, or explode a mixture of oxygen and hydrogen. Though sonorous waves pass in each case through the air, the candle is absolutely insensible to the sound; there is no motion of the flame.

I now urge from this small blow-pipe a narrow stream of air through the flame of the candle, producing thereby an incipient flutter, and reducing at the same time the brightness of the flame. When I now sound a whistle, the flame jumps visibly. The experiment may be so arranged that when the whistle sounds, the flame shall be either restored almost to its pristine brightness, or that the amount of light it still possesses shall disappear.

The blow-pipe flame of our laboratory is totally un-

FIG. 115.

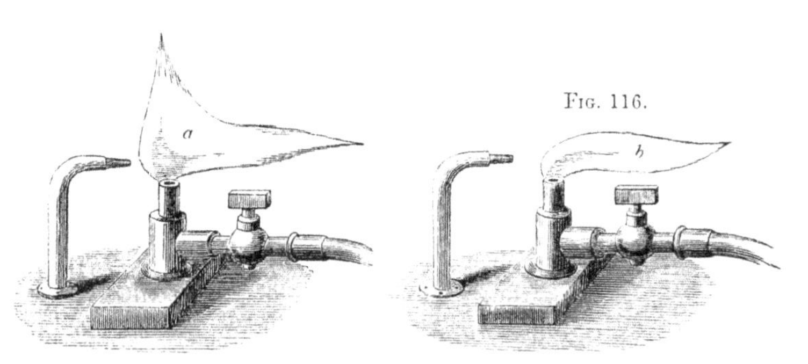

FIG. 116.

affected by the sound of the whistle as long as no air is urged through it. By properly tempering the force of the blast I obtain a flame of the shape shown in fig. 115, the blast not being sufficiently powerful to urge the whole of the flame forwards. On sounding the whistle the erect portion of the flame drops down, and while it continues to sound we have a flame of the form shown in fig. 116.

Here moreover is a fish-tail flame, which burns brightly and steadily, refusing to respond to any sound, musical or unmusical. I urge against the broad face of the flame a stream of air from a blow-pipe. The flame is cut in two by the air, and now, when the whistle is sounded, it instantly starts. A knock on the table causes the two half-flames to unite, and form, for an instant, a single flame of the ordinary shape. By a slight variation of the experiment, the two side-flames disappear when the whistle is sounded, a central luminous tongue being thrust forth in their stead.

Before you now is another thin sheet of flame, also issuing from a common fish-tail burner, fig. 117. You might

Fig. 118.

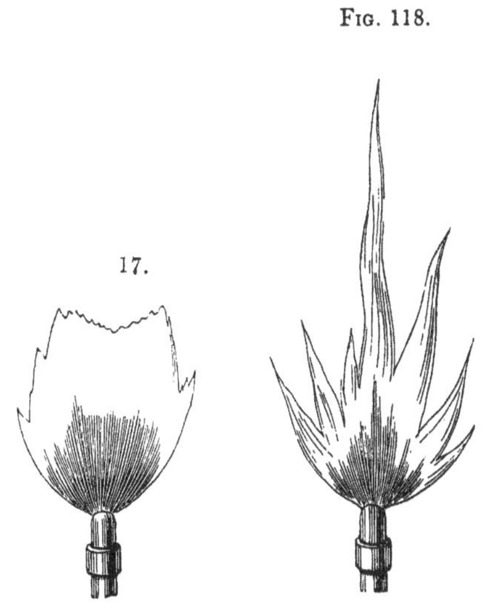

17.

sing to it, varying the pitch of your voice, no shiver of the flame would be visible. You might employ pitch-pipes, tuning-forks, bells, and trumpets, with a like absence of all effect. A barely perceptible motion of the interior of the flame may be noticed when this shrill whistle is blown close to it. By turning the cock more fully on I bring the flame to the verge of flaring. And now, when the whistle is blown, you see an extraordinary appearance. The flame thrusts out seven quivering tongues, fig. 118. As long as the sound continues the tongues jut forth, being violently agitated; the moment the sound ceases, the tongues disappear, and the flame becomes quiescent.

Passing from a fish-tail to a bat's-wing burner, we obtain this broad, steady flame, fig. 119. It is quite insensible to the loudest sound which would be tolerable here. The flame is fed from this small gas-holder,* which places a greater pressure at my disposal than that existing in the pipes of the Institution. I enlarge the flame and now a slight flutter of its edge answers to the sound of the whistle.

Fig. 120.

Fig. 119.

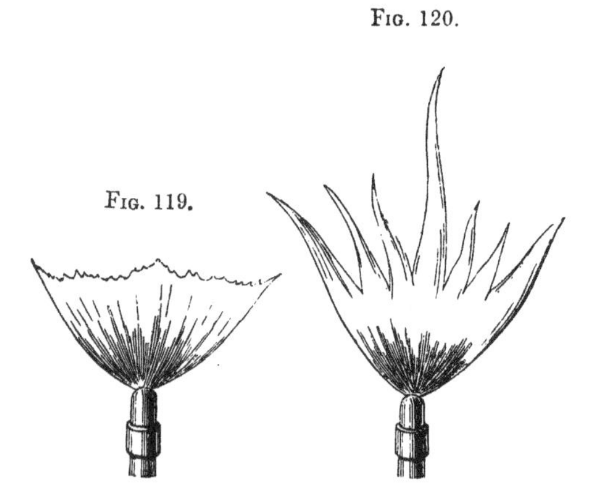

Finally I turn on gas until the flame is on the point of roaring, as flames do when the pressure is too great. I now sound the whistle; the flame roars, and suddenly assumes the form shown in fig. 120.

* A gas-bag properly weighted also answers for these experiments.

When I strike a distant anvil with a hammer, the flame instantly responds by thrusting forth its tongues.

An essential condition to entire success in these experiments disclosed itself in the following manner. I was in a room illuminated by two fish-tail flames. One of them jumped to a whistle, the other did not. The gas of the non-sensitive flame was turned off, additional pressure being thereby thrown upon the other flame; it flared, and its cock was turned so as to lower the flame. It now proved non-sensitive, however close it might be brought to the point of flaring. The narrow orifice of the half-turned cock appeared to interfere with the action of the sound. When the gas was turned fully on, the flame being lowered by opening the cock of the second burner, it became again sensitive. Up to this time a great number of burners had been tried, including some with single orifices, but, with many of them, the action was *nil*. Acting, however, upon the hint conveyed by this observation, the pipes which fed the flames were widely opened; the consequence was, that our most refractory burners were thus rendered sensitive.

The observation of Dr. Leconte is thus easily and strikingly illustrated; in our subsequent, and far more delicate experiments, the precaution just referred to is still more essential.

Mr. Barrett, late laboratory assistant in this place, first observed the shortening of a tall flame issuing from the single orifice of this old burner, when the higher notes of a circular plate were sounded; and, by the selection of more suitable burners, he afterwards succeeded in rendering the flame extremely sensitive.* Observing the precaution above adverted to, we can readily obtain, in an exalted degree, the shortening of the flame. It is

* For Mr. Barrett's own account of his experiments I refer the reader to the Philosophical Magazine for March, 1867.

now before you, being 18 inches long and smoking copiously. When I sound the whistle the flame falls to a height of 9 inches, the smoke disappearing, and the flame increasing in brightness.

A long flame may be shortened and a short one lengthened, according to circumstances, by these sonorous vibrations. Here, for example, are two flames, issuing from rough burners formed from pewter tubing. The one flame, fig. 121, is long, straight, and smoky; the other, fig. 122,

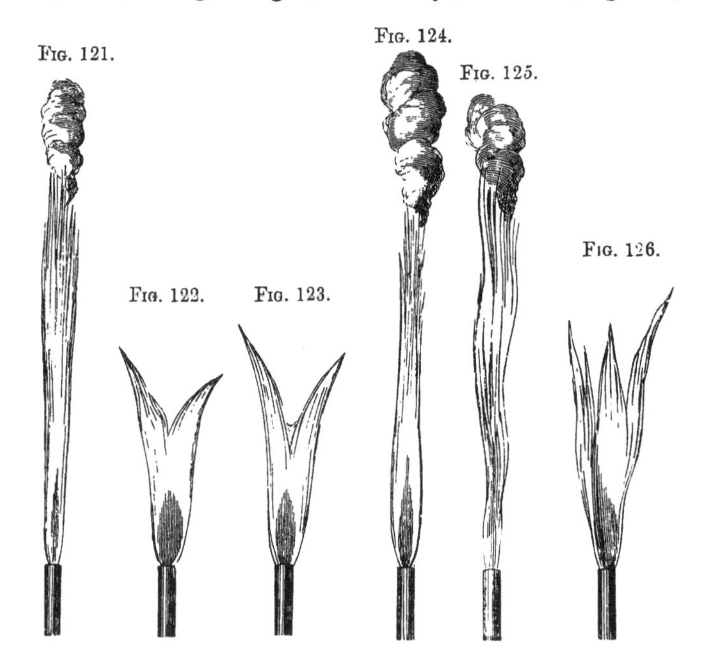

Fig. 121. Fig. 122. Fig. 123. Fig. 124. Fig. 125. Fig. 126.

is short, forked, and brilliant. On sounding the whistle, the long flame becomes short, forked, and brilliant, as in fig. 123; the forked flame becomes long and smoky, as in fig. 124. As regards, therefore, their response to the sound of the whistle, one of these flames is the complement of the other.

In fig. 125 is represented another smoky flame which, when the whistle sounds, breaks up into the form shown in fig. 126.

The foregoing experiments illustrate the lengthening and shortening of flames by sonorous vibrations. They are also able to produce *rotation*. We have here several home-made burners, from which issue flat flames, each about ten inches high, and three inches across at their widest part. The burners are purposely so formed that the flames are dumpy and forked. When the whistle sounds, the plane of each flame turns ninety degrees round, and continues in its new position as long as the sound continues.

A flame of admirable steadiness and brilliancy now burns before you. It issues from a single circular orifice in a common iron nipple. This burner, which requires great pressure to make its flame flare, has been specially chosen for the purpose of enabling you to observe, with distinctness, the gradual change from apathy to sensitiveness. The flame is now 4 inches high, and is quite indifferent to sound. By increasing the pressure I make its height 6 inches; it is still indifferent. I make it 12 inches; a barely perceptive quiver responds to the whistle. I make it 16 or 17 inches high, and now it jumps briskly the moment an anvil is tapped or the whistle sounded. I augment the pressure, the flame is now 20 inches long, and you observe a quivering at intervals, which announces that it is near roaring. A slight increase of pressure causes it to roar, and shorten at the same time to 8 inches. I diminish the pressure a little; the flame is again 20 inches long, but it is on the point of roaring and shortening. Like the singing-flames which were started by the voice, it stands on the brink of a precipice. The proper sound pushes it over. It shortens when the whistle sounds, exactly as it did when the pressure was in excess.

The action reminds one of the story of the Swiss muleteers, who are said to tie up their bells at certain places lest the tinkle should bring an avalanche down. The snow must be very delicately poised before this could occur. I believe it never did occur, but our flame illustrates the principle. We bring it to the verge of falling, and the sonorous pulses precipitate what was already imminent. This is the simple philosophy of all these sensitive flames.

When the flame flares, the gas in the orifice of the burner has been thrown into vibration; conversely, when the gas in the orifice is thrown into vibration, the flame, if sufficiently near the flaring point, will flare. Thus the sonorous vibrations, by acting on the gas in the passage of the burner, become equivalent to an augmentation of pressure in the holder. In fact, we have here revealed to us the physical cause of flaring through excess of pressure, which, common as it is, has never, I believe, been hitherto explained. The gas encounters friction in the orifice of the burner, which, when the force of transfer is sufficiently great, throws the issuing stream into the state of vibration that produces flaring. It is because the flaring is thus caused, that an almost infinitely small amount of energy in the form of vibrations of the proper period, can produce an effect equivalent to a considerable increase of pressure. Augmentation of pressure is, in fact, a clumsy means of causing a flame to flare.

All sounds are not equally effective on the flame; waves of special periods are required to produce the maximum effect. The effectual periods are those which synchronise with the waves produced by the friction of the gas itself against the sides of its orifice. With some of the flames which you have already seen, a low, deep whistle is more effective than a shrill one. With the flame now before you, the exciting tremors must be very rapid, and the sound consequently shrill. I have here a tuning-fork which vibrates 256 times in a second, emitting a clear and forcible note. It has no effect upon this flame. Here also are three other forks, vibrating respectively 320, 384, and 512 times in a second. Not one of them produces the slightest impression upon the flame. But, besides their fundamental tones, these forks, as you know, can be caused to yield a series of overtones of very high pitch. I sound this series: the vibrations are now 1,600, 2,000, 2,400, and 3,200 per second, respectively. The flame jumps in response to each of these sounds; the response to the highest sound of the series being the most prompt and energetic of all.

To the tap of a hammer upon a board the flame responds; but to the tap of the same hammer upon an anvil the response is much more brisk and animated. The reason is, that the clang of the anvil is rich in the higher tones to which the flame is most sensitive.

The powerful tone obtained when our inverted bell is reinforced by its resonant tube has no power over this flame. The bell is now sounding, but the flame is unmoved. I bring a halfpenny into contact with the vibrating surface the consequent rattle contains the high notes to which the flame is sensitive. It instantly shortens, flutters, and roars when the coin touches the bell. I hold in my hand a smaller bell, the hammer of which is worked by clock-work. I send my assistant to the most distant part of the gallery, where he detaches the hammer. The strokes follow each other in rhythmic succession, and at every stroke the flame falls from a height of 20 to a height of 8 inches, roaring as it falls.

The rapidity with which sound is propagated through air is well illustrated by these experiments. There is no sensible interval between the stroke of the bell and the ducking of the flame.

When the sound acting on the flame is of very short duration a curious and instructive effect is observed. The sides of the flame half-way down, and lower, are seen suddenly fringed by luminous tongues, the central flame remaining apparently undisturbed in both height and thickness. The flame in its normal state is shown in fig. 127, and with its fringes in fig. 128. The effect is due to the retention of the impression upon the retina. The flame actually falls as low as the fringes, but its recovery is so quick that to the eye it does not appear to shorten at all.*

FIG. 127. FIG. 128.

The most marvellous flame hitherto discovered is now before you. It issues from the single orifice of a steatite burner, and reaches a height of 24 inches. The slightest tap on a distant anvil reduces its height to 7 inches. When I shake this bunch of keys the flame is violently agitated, and emits a loud roar. The dropping of a sixpence into a hand already containing coin, at a distance of 20 yards, knocks the flame down. I cannot walk across the floor without agitating the flame. The creaking of my boots sets it in violent commotion. The crumpling, or tearing of a bit of paper, or the rustle of a silk dress, does the same. It is startled by the patter of a raindrop. I hold a watch near the flame; nobody hears its ticks; but you all see their effect upon the flame. At every tick it falls. The winding up of the watch also produces tumult. The

* Numerous modifications of these experiments are possible. Other inflammable gases than coal gas may be employed. Mixtures of gases have also been found to yield beautiful and striking results. An infinitesimal amount of mechanical impurity has been found to exert a powerful influence.

twitter of a distant sparrow shakes the flame down; the note of a cricket would do the same. From a distance of 30 yards I have chirruped to this flame, and caused it to fall and roar. I repeat a passage from Spenser:—

FIG. 129.

b

a

x

> Her ivory forehead full of bounty brave,
> Like a broad table did itself dispread;
> For love his lofty triumphs to engrave,
> And write the battles of his great godhead.
> All truth and goodness might therein be read,
> For there their dwelling was, and when she spake,
> Sweet words, like dropping honey she did shed;
> And through the pearls and rubies softly brake
> A silver sound, which heavenly music seemed to make.

The flame picks out certain sounds from my utterance; it notices some by the slightest nod, to others it bows more distinctly, to some its obeisance is very profound, while to many sounds it turns an entirely deaf ear.

In fig. 129, this tall, straight, and brilliant flame is represented. On chirruping to it, or on shaking a bunch of keys within a few yards of it, it falls to the size shown in fig. 130, the whole length, *a b*, of the flame being suddenly abolished. The light at the same time is practically destroyed, a pale and almost non-luminous residue of it alone remaining. These figures are taken from photographs of the flame.

FIG. 130.

y

In our experiments downstairs we have called this the 'vowel flame,' because the different vowel sounds affect it differently. We have already learned how these sounds are formed; that they differ from each other through the admixture of higher tones with the fundamental one. It is to these tones, and not to the fundamental one, that our flame is sensitive. I utter a loud and sonorous U, the flame remains steady; I change the sound to O, the flame quivers; I sound E, and now the flame is strongly affected. I utter the words *boot*, *boat*, and *beat* in succession. To the first there is no response; to the second, the flame starts; but by the third it is thrown into greater commotion; the sound *Ah!* is still more powerful. Did we not know the constitution of vowel sounds this deportment would be an insoluble enigma. As it is, however, the flame is a demonstrator of the theory of vowel sounds. It is most sensitive to sounds of high pitch; hence we should infer that the sound *Ah!* contains higher notes than the sound E; that E contains higher notes than O; and O higher notes than U. I need not say that this agrees perfectly with the analysis of Helmholtz.

This flame is peculiarly sensitive to the utterance of the letter S. If the most distant person in the room were to favour me with a 'hiss,' the flame would instantly sympathise with him. A hiss contains the elements that most forcibly affect this flame. The gas issues from its burner with a hiss, and an external sound of this character is therefore exceedingly effective. I hold in my hand a metal box, containing compressed air. I turn the cock for a moment, so as to allow a puff to escape—the flame instantly ducks down, not by any transfer of air from the box to the flame, for I stand at a distance which utterly excludes this idea; it is the *sound* that affects the flame. I send a man to the most distant part of the gallery, where he permits the compressed air to issue in puffs from the box; at every puff the flame suddenly falls. Thus the hiss of the issuing air at the one orifice precipitates the tumult of the flame at the other.

Finally, I place this musical box upon the table, and permit it to play. The flame behaves like a sentient creature; bowing slightly to some tones, but courtesying deeply to others.

I at one time intended to approach this subject of sensitive flames through a series of experiments which, had the flames not been seen, would have appeared more striking than I can expect them to be now. It is not to the flame, as such, that we owe the extraordinary phenomena which have been just described. Effects substantially the same are obtained when a jet of unignited gas, of carbonic acid, hydrogen, or even air itself, issues from an orifice under proper pressure. None of these gases, however, can be seen in its passage through air; and, therefore, we must associate with them some substance which, while sharing their motions, will reveal those motions to the eye. The method employed from time to time in this place, of rendering aërial vortices visible is well known to many of you. By tapping a membrane which closes the broad mouth of a large funnel filled with smoke, we obtain beautiful smoke rings, which reveal the motion of the air. By associating smoke with our gas-jets, in the present instance, we can also trace their course, and when this is done, the unignited gas proves as sensitive as the flames. The smoke-jets jump, they shorten, they split into forks, or lengthen into columns, when the proper notes are sounded. The experiments are made in this way. Underneath a gasometer are placed two small basins, the one containing hydrochloric acid and the other ammonia. Fumes of sal-ammoniac are thus copiously formed, and mingle with the gas contained in the holder. We may, as already stated, operate with coal gas, carbonic acid, air, or hydrogen: each of them yields good effects. Here also our excellent steatite burner maintains that supremacy which it exhibited with the flames. From it I can cause to issue a thin column of smoke. On sounding the whistle, which was so effective with the flames, it is found ineffective. When, moreover, the highest notes of a series of Pandean pipes are sounded, they are also ineffective. Nor will the lowest notes answer. But when a certain pipe, which stands about the middle of the series, is sounded, the smoke-column falls, forming a short stem with a thick, bushy head. It is also pressed down, as if by a vertical wind, by a knock upon the table. At every tap, it drops. A stroke on an anvil, on the contrary, produces little or no effect. In fact, the notes here effective are of a much lower pitch than those which were most efficient in the case of the flames.

The amount of shrinkage exhibited by some of these smoke columns, in proportion to their length, is far greater than that of the flames. A tap on the table causes a smoke-jet eighteen inches high, to shorten to a bushy bouquet, with a stem not more than an inch in height.

Fig. 131.

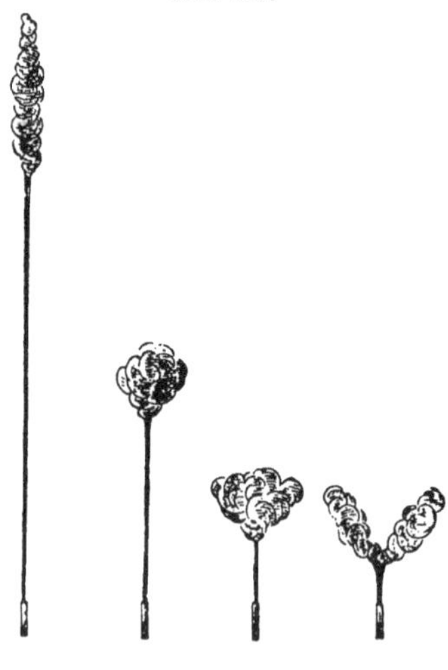

The smoke column, moreover, responds to the voice. A cough knocks it down; and it dances to the tune of a musical box. Some notes cause the mere top of the smoke-column to gather itself up into a bouquet. At other notes the bouquet is formed midway down; while notes of more suitable pitch cause the column to contract itself to a cumulus not much more than an inch above the end of the burner. As the music continues, the action of the smoke-column consists of a series of rapid leaps from one of these forms to another. Various forms of the dancing jet are shown in fig. 131.

In a perfectly still atmosphere these slender smoke-columns rise sometimes to a height of nearly two feet, apparently vanishing into air at the summit. When this is the case our most sensitive flames fall far behind them in delicacy; and though less striking than the flames, the smoke-wreaths are often more graceful. Not only special words, but every word, and every syllable of the foregoing stanza from Spenser tumbles a really sensitive smoke-jet into confusion. To produce such effects, a perfectly tranquil atmosphere is necessary. Flame experiments, in fact, are possible in an atmosphere where smoke-jets are utterly unmanageable.*

We have thus far confined our attention to jets of ignited and unignited coal-gas; of carbonic acid, hydrogen, and air. We will now turn to jets of water. And here a series of experiments, remarkable for their beauty, has long existed, which claim relationship to those just described. These are the experiments of Felix Savart on liquid veins; which have been repeated, verified, and modified, in various ways in this place. If the bottom of a vessel containing water be pierced by a circular orifice, the descending liquid vein will exhibit two parts which are unmistakably distinct. The part of the vein nearest the orifice is steady and limpid, presenting almost the appearance of a solid glass rod. It decreases in diameter as it descends, reaches a point of maximum contraction, from which point downwards it appears turbid and unsteady. The course of the vein, moreover, is marked by periodic swellings and contractions. Savart has represented the vein in the manner shown in fig. 132. In this figure, *a* is the orifice end of the vein, the part *a n* is limpid and steady, while all the part below *n* is in a state of quivering motion. This lower part of the vein appears continuous to the eye; still, when the finger is passed rapidly across it, it is some-

* Could the jets of unignited gas be seen without any admixture of smoke, their sensitiveness I doubt not might be increased.

FIG. 132. FIG. 133.  FIG. 134.

times not wetted. This, of course, could not be the case if the vein were really continuous. The upper portion of the vein, moreover, intercepts vision; the lower portion, even when the liquid is mercury, does not. In fact, the vein resolves itself, at *n*, into liquid spherules, its apparent continuity being due to the retention of the impressions made by the falling drops upon the retina. If the drops succeed each other in intervals of a tenth of a second or less, then, before the impression made by any drop vanishes, it is renewed by its successor, and no rupture of continuity can be observed. If, while looking at the disturbed portion of the vein, the head be suddenly lowered, the descending column will be resolved for a moment into separate drops. Perhaps the simplest way of reducing the vein to its constituent spherules is one long ago adopted by myself,—namely, to illuminate the vein, in a dark room, by a succession of electric flashes. Every flash reveals the drops, as if they were perfectly motionless in the air.

Could the appearance of the vein illuminated by a single flash be rendered permanent, it would be that represented in fig. 133. And here we find revealed the cause of those swellings and contractions which the disturbed portion of the vein exhibits. The drops, as they descend, are continually changing their forms. When first detached from the end of the limpid portion of the vein the drop is a prolate spheroid, with its longest axis vertical. But a liquid cannot retain this shape, if abandoned to the forces of its own molecules. The spheroid seeks to become a sphere. The longer diameter therefore shortens; but, like a pendulum, which seeks to return to its position of rest, the contraction of the vertical diameter goes too far, and the drop becomes an oblate spheroid. Now the contractions of the jet are formed at those places where the longest axis of the drop is vertical, while the swellings appear where the longest axis is horizontal. It will be noticed, that between every two of the larger drops is a third of much smaller dimensions. Whenever a large drop is detached, by a kind of kick on the part of the retreating vein, a little satellite is shaken after it. According to Savart, their appearance is invariable.

This breaking up of a liquid vein into drops has been a subject of much discussion. I hold it to be due to the tremors imparted to the vein by its friction against the boundaries of its orifice. To this point Savart traced its pulsations, though he did not think that friction was their cause. Whatever their cause may be, the pulsations exist, and they are powerfully influenced by sonorous vibrations, which render the limpid portion of the vein shorter than it would otherwise be. In the midst of a large city it is hardly possible to obtain the requisite aërial tranquillity for the full development of the continuous portion of the vein; still, Savart was so far able to withdraw his vein from the influence of such irregular vibrations, that its limpid portion became elongated to the extent shown in fig. 134. Fig. 132, it will be understood, represents the vein exposed to the irregular vibrations of the city of Paris, while fig. 134 represents a vein produced under precisely the same conditions, but withdrawn from those vibrations.

The drops into which the vein finally resolves itself, are incipient even in its limpid portion, announcing themselves there as annular protuberances, which become more and more pronounced, until finally they separate. Their birth-place is the orifice itself, and under even moderate pressure they succeed each other with sufficient rapidity to produce a feeble musical note. By permitting the drops to fall upon a membrane, the pitch of this note may be fixed; and now we come to the point which connects the phenomena of liquid veins with those of sensitive flames and smoke-jets. If a note in unison with that of the vein be sounded near it, the limpid portion instantly shortens; the pitch may vary to some extent, and still cause a shortening; but the unisonant note is the most effectual. Savart's experiments on vertically descending veins have been recently repeated in our laboratory with striking effect. From a distance of 30 yards the limpid portion of the vein has been shortened by the sound of an organ-pipe of moderate intensity but of the proper pitch.

The excellent French experimenter also caused his vein to issue horizontally and at various inclinations to the horizon, and found that, in certain cases, sonorous vibrations were competent to cause a jet to divide into two or three branches. In these experiments, the liquid was permitted to issue through an orifice in a thin plate. Instead of this, however, we will resort to our favourite steatite burner; for with water also it asserts the same mastery over its fellows that it exhibited with flames and smoke-jets. It will, moreover, reveal to us some entirely novel results. By means of an india-rubber tube the burner is connected with the water pipes of the Institution, and, by pointing it obliquely upwards, we obtain a fine parabolic jet, fig. 135. At a cer-

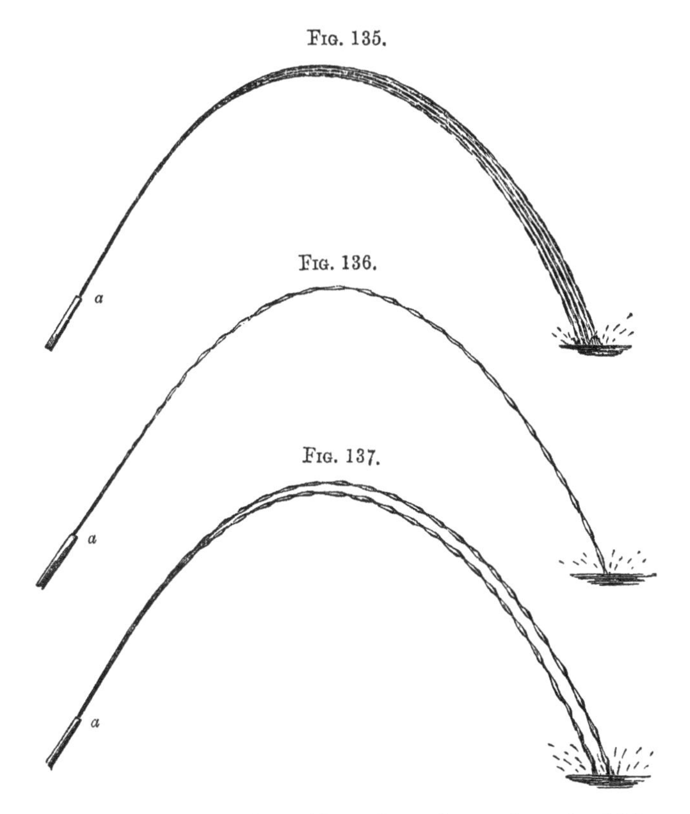

Fig. 135.

Fig. 136.

Fig. 137.

tain distance from the orifice, the vein resolves itself into beautiful spherules, whose motions are not rapid enough to make the vein appear continuous. At the vertex of the parabola the spray of pearls is more than an inch in width, and, further on, the drops are still more widely scattered. On sweeping a fiddle-bow across a tuning-fork, which executes 512 vibrations in a second: the scattered drops, as if drawn together by their mutual attractions, instantly close up and form an apparently continuous liquid arch some feet in height and span, fig. 136. As long as the proper note is maintained the vein looks like a frozen band, so motionless does it appear. I stop the fork, and now the arch is shaken asunder, and we have the same play of liquid pearls as before. Every sweep of the bow, however, causes the drops to fall into a common line of march.

A pitch-pipe, or an organ-pipe yielding the note of this tuning-fork also powerfully controls the vein. My voice does the same. On pitching it to a note of moderate intensity, I cause the wandering drops to gather themselves together. At a distance of 20 yards, my voice is, to all appearance, as powerful in curbing the vein, and causing its drops to close up, as when I stand close to the issuing jet.

The effect of beats upon the vein is also beautiful and instructive. They may be produced either by organ-pipes or by tuning-forks. Before you are two forks, one of which vibrates 512 times, and the other 508 times in a second. You will learn, in our next lecture, that when these two forks are sounded together we ought to have four beats in a second. I sound the two forks, and find that the liquid vein gathers up its pearls, and scatters them in synchronism with the beats. When standing near the vein we notice a rhythmic movement of the spots of light reflected from it, keeping time with the beats The alternate retreat and advance of the point where the drops are first formed is executed in the same period, and is very beautiful. The sensitiveness of this vein is now astounding; it rivals that of the ear itself. Placing the two tuning-forks on a distant table, and permitting the beats to die gradually out, the vein continues its rhythm almost as long as hearing is possible. A more sensitive vein might actually prove superior to the ear—a very surprising result, considering the marvellous delicacy of this organ.*

By introducing a Leyden jar into the circuit of a powerful induction coil, I obtain, as those acquainted with the coil well know, a series of dense and dazzling flashes of light, each of momentary duration. I darken the room, and, with a succession of flashes, illuminate the vein. Its drops are rendered distinct, every one of them being transformed into a little star of intense brilliancy. They are scattered widely apart. I call to the jet in the proper tone of voice. It instantly gathers its drops together into a necklace of inimitable beauty. I suspend my voice, the vein goes to pieces; I call again, and the straggling stars arrange themselves once more in succession along the curve. While thus arranged, I gently shake the india-rubber tube which feeds the jet, and obtain interlacing strings of luminous pearls.

In these experiments the whole vein gathers itself into a single arched band when the proper note is sounded; but, by varying the experiment, it may be caused to divide into two or more such bands, as shown in fig. 137. Drawings, however, are ineffectual here; for the wonder of these experiments depends mainly on the sudden transition of the vein from one state to the other. In the *motion* dwells the surprise, and this no drawing can render.†

* When these two tuning-forks were placed *in contact* with a vessel from which a liquid vein issued, the visible action on the vein continued long after the forks had ceased to be heard.

† The experiments on sounding flames have been recently considerably extended by my assistant. By causing flame to rub against flame, various musical sounds can be obtained, some resembling those of a trumpet, some those of a lark. By the friction of unignited gas-jets, similar though less intense effects are produced. When the two flames of a fish-tail burner are permitted to impinge upon a plate of platinum, as in Scholl's 'perfectors,' the sounds are trumpet-like, and very loud.

SUMMARY OF LECTURE VI.

When a gas-flame is placed in a tube, the air in passing over the flame is thrown into vibration, musical sounds being the consequence.

Making allowance for the high temperature of the column of air associated with the flame, the pitch of the note is that of an open organ-pipe of the length of the tube surrounding the flame.

The vibrations of the flame, while the sound continues, consist of a series of periodic extinctions, total or partial; between every two of which the flame partially recovers its brightness.

The periodicity of the phenomenon may be demonstrated by means of a concave mirror which forms an image of the vibrating flame upon a screen. When the image is sharply defined, the rotation of the mirror reduces the single image to a series of separate images of the flame. The dark spaces between the images correspond to the extinctions of the flame, while the images themselves correspond to its periods of recovery.

Besides the fundamental note of the associated tube, the flame can also be caused to excite the higher harmonics of the tube. The successive divisions of the column of air are those of an open organ-pipe when its harmonic tones are sounded.

On sounding a note nearly in unison with a tube containing a silent flame, the flame jumps, and if the position of the flame in the tube be rightly chosen, the extraneous sound will cause the flame to sing.

While the flame is singing a note nearly in unison with its own produces beats; and the flame is seen to jump in synchronism with the beats. The jumping is also observed when the position of the flame within its tube is not such as to enable it to sing.

NAKED FLAMES.

When the pressure of the gas which feeds a naked flame is augmented, the flame, up to a certain point, increases in size. But if the pressure be too great, the flame roars or flares.

The roaring or flaring of the flame is caused by the state of vibration into which the gas is thrown in the orifice of the burner, when the pressure which urges it through the orifice is excessive.

If the vibrations in the orifice of the burner be superinduced by an extraneous sound, the flame will flare under a pressure less than that which, of itself, would produce flaring.

The gas under excessive pressure has vibrations of a definite period impressed upon it as it passes through the burner. To operate with a maximum effect upon the flame the external sound must contain vibrations synchronous with those of the issuing gas.

When such a sound is chosen, and when the flame is brought sufficiently near its flaring point, it furnishes an acoustic reagent of unexampled delicacy.

At a distance of 30 yards, for example, the chirrup of a house sparrow would be competent to throw the flame into commotion.

It is not to the flame, as such, that we are to ascribe these effects. Effects substantially similar are produced when we employ jets of unignited coal-gas, carbonic acid, hydrogen, or air. These jets may be rendered visible by smoke, and the smoke jets show a sensitiveness to sonorous vibrations even greater than that of the flames.

When a brilliant sensitive flame illuminates an otherwise dark room, in which a suitable bell is caused to strike, a series of periodic quenchings of the light by the sound occurs. Every stroke of the bell is accompanied by a momentary darkening of the room.

Savart's experiments on the influence of sonorous vibrations on jets of water belong to the same class of effects. This subject is treated with sufficient fulness in the foregoing lecture, and still almost as briefly as a summary.

LECTURE VII.

LAW OF VIBRATORY MOTIONS IN WATER AND AIR—SUPERPOSITION OF VIBRATIONS—INTERFERENCE AND COINCIDENCE OF SONOROUS WAVES—DESTRUCTION OF SOUND BY SOUND—COMBINED ACTION OF TWO SOUNDS NEARLY IN UNISON WITH EACH OTHER—THEORY OF BEATS—OPTICAL ILLUSTRATION OF THE PRINCIPLE OF INTERFERENCE—AUGMENTATION OF INTENSITY BY PARTIAL EXTINCTION OF VIBRATIONS—RESULTANT TONES—CONDITIONS OF THEIR PRODUCTION — EXPERIMENTAL ILLUSTRATIONS — DIFFERENCE TONES AND SUMMATION TONES—THEORIES OF YOUNG AND HELMHOLTZ.

FROM a boat in Cowes harbour, in moderate weather, I have often watched the masts and ropes of the ships, as mirrored in the water. The images of the ropes revealed the condition of the surface, indicating by long and wide protuberances the passage of the larger rollers, and, by smaller indentations, the ripples which crept like parasites over the sides of the nobler waves. The sea was able to accommodate itself to the requirements of all its undulations, great and small. When I touched the surface with my oar, or permitted the drops to fall from the oar into the water, there was also room for the tiny wavelets thus generated. This carving of the surface by waves and ripples had its limit only in my powers of observation; every wave and every ripple asserted its right of place, and retained its individual existence, amid the crowd of other motions which agitated the water.

The law that rules this chasing of the sea, this crossing and intermingling of innumerable small waves, is *that the resultant motion of every particle of water is the sum of the individual motions imparted to it.* If any particle be acted on at the same moment by two impulses, both of which tend to raise it, it will be lifted by a force equal to the sum of both. If acted upon by two impulses, one of which tends to raise it, and the other to depress it, it will be acted upon by a force equal to the difference of both. When, therefore, I speak of the sum of the motions, I mean the *algebraic sum*, regarding the motions which tend to raise the particle as positive, and those which tend to depress it as negative.

When two stones are cast into smooth water, 20 or 30 feet apart, round each stone is formed a series of expanding circular waves, every one of which consists of a ridge and a furrow. The waves at length touch, and then cross each other, carving the surface into little eminences and depressions. Where ridge coincides with ridge, we have the water raised to a double height; where furrow coincides with furrow, we have it depressed to a double depth. Where ridge coincides with furrow, we have the water reduced to its average level. The resultant motion of the water at every point is, as above stated, the algebraic sum of the motions impressed upon that point. And if, instead of two sources of disturbance, we had ten, or a hundred, or a thousand, the consequence would be the same; the actual result might transcend our powers of observation, but the law above enunciated would still hold good.

Instead of the intersection of waves from two distinct centres of disturbance, we may cause direct and reflected waves, from the same centre, to cross each other. Many of you know the beauty of the effects observed when the light reflected from ripples of water, contained in a common tray, is received upon our screen. When mercury is employed the effect is more brilliant still. Here, by a proper mode of agitation, direct and reflected waves may be caused to cross and interlace, and by the most wonderful self-analysis to untie their knotted scrolls. The adjacent figure, fig. 138, which is copied from the work

FIG. 138.

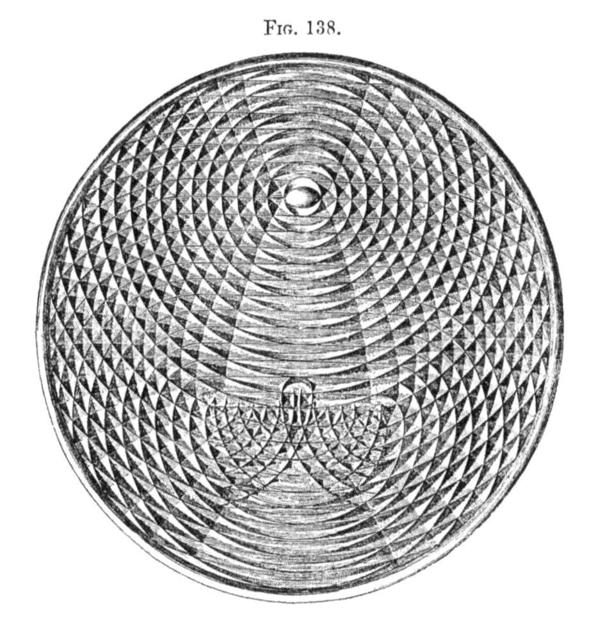

of the brothers Weber, will give some idea of the beauty of these effects. It represents the chasing produced by the intersection of direct and reflected water-waves in a circular vessel, the point of disturbance, marked by the smallest circle in the figure, being midway between the centre and the circumference.

This power of water to accept and transmit multitudinous impulses is shared by air, which concedes the right of space and motion to any number of sonorous waves. The same air is competent to accept and transmit the vibrations of a thousand instruments at the same time. When we try to visualise the motion of that air—to present to the eye of the mind the battling of the pulses direct and reverberated

—the imagination retires baffled from the attempt. Still, amid all the complexity, the law above enunciated holds good, every particle of air being animated by a resultant motion, which is the algebraic sum of all the individual motions imparted to it. And the most wonderful thing of all is, that the human ear, though acted on only by a cylinder of that air, which does not exceed the thickness of a quill, can detect the components of the motion, and, aided by an act of attention, can even isolate from the aërial entanglement any particular sound.

I draw my bow across a tuning-fork, which for distinction's sake I will call A, and cause it to send a series of sonorous waves through the air. I now place a second fork, B, behind the first, and throw it also into vibration. From B waves issue which pass through the air already traversed by the waves from A. It is easy to see that the forks may so vibrate that the condensations of the one shall coincide with the condensations of the other, and the rarefactions of the one with the rarefactions of the other. If this be the case the two forks will assist each other. The condensations will, in fact, become more condensed, the rarefactions more rarefied, and as it is upon the difference of density between the condensations and rarefactions that loudness depends, the two vibrating forks, thus supporting each other, will produce a sound of greater intensity than that of either of them vibrating alone.

It is, however, also easy to see that the two forks may be so related to each other that one of them shall require a condensation at the place where the other requires a rarefaction; that one fork, for example, shall urge the air-particles forward, while the other urges them backward. If the opposing forces be equal, particles so solicited will move neither backwards nor forwards, and the aërial rest which corresponds to silence is the result. Thus, it is possible, by adding the sound of one fork to that of another, to abolish the sounds of both. We have here a phenomenon which, above all others, characterises wave-motion. It was this phenomenon, as manifested in optics, that led to the undulatory theory of light, the most cogent proof of that theory being based upon the fact that, by adding light to light, we may produce darkness, just as we can produce silence by adding sound to sound.

During the vibration of a tuning-fork the distance between its two prongs is alternately increased and diminished. Let us call the motion which increases the distance the *outward swing*, and that which diminishes the distance the *inward swing* of the fork. And let us suppose that our two forks, A and B, reach the limits of their outward swing and their inward swing at the same moment. In this case the *phases* of their motion, to use the technical term, are the same. For the sake of simplicity we will confine our attention to the right-hand prongs, A and B, fig. 139, of the two forks, neglecting

Fig. 139.

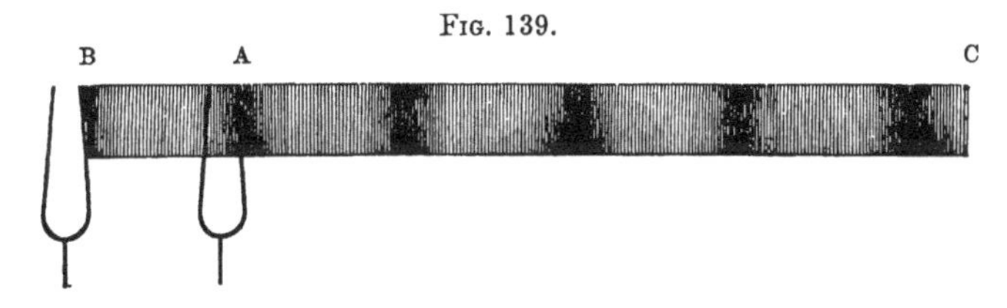

the other two prongs; and now let us ask what must be the distance between the prongs A and B, when the condensations and rarefactions of both, indicated respectively by the dark and light shading, coincide? A little reflection will make it clear, that if the distance from B to A be equal to the length of a whole sonorous wave, coincidence between the two systems of waves must follow. The same would evidently occur where the distance between A and B is two wave-lengths, three wave-lengths, four wave-lengths—in short, any number of whole wave-lengths. In all such cases we should have *coincidence* of the two systems of waves, and consequently a reinforcement of the sound of the one fork by that of the other. Both the condensations and rarefactions between A and C are, in this case, more pronounced than they would be if either of the forks were suppressed.

But if the prong B be only half the length of a wave behind A, what must occur? Manifestly the rarefactions of one of the systems of waves will then coincide with the condensations of the other system, and we shall have *interference*; the air to the right of A being reduced to quiescence. This is shown in fig. 140, where the uniformity

FIG. 140.

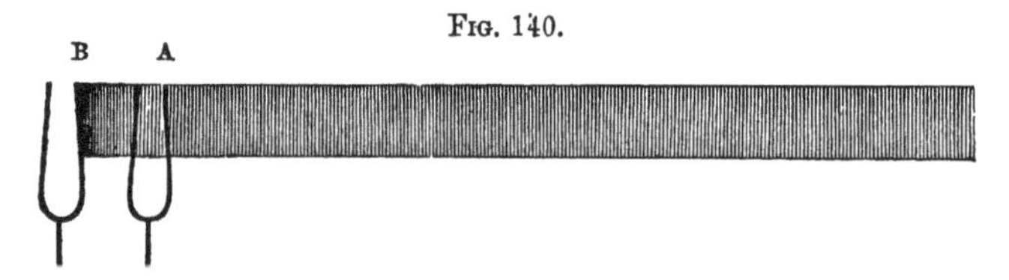

of tint indicates an absence both of condensations and rarefactions. When B is two half wave-lengths behind A, we have, as already explained, coincidence; when it is three half wave-lengths distant, we have again interference. Or, expressed generally, we have coincidence or interference according as the distance between the two prongs amounts to an even or an odd number of semi-undulations. Precisely the same is true of the waves of light. If through any cause one system of ethereal waves be any *even* number of semi-undulations behind another system, the two systems support each other when they coalesce, and we have more light. If the one system be any *odd* number of semi-undulations behind the other, they interfere with each other, and a destruction of light is the result of their coalescence.

Sir John Herschel first proposed to divide a stream of

sound into two branches, of different lengths, causing the branches afterwards to reunite, and to interfere with each other. This idea has been recently followed out with success by M. Quincke; and it has been still further improved upon by M. König. The principle of these experiments will be at once evident from fig. 141. The tube *o f* divides into two branches at *f*, the one branch being carried round *n*, and the other round *m*. The two branches are caused to reunite at *g*, and to end in a common canal, *g p*. The portion, *b n*, of the tube which slides over *a b*, can be drawn out as shown in the figure, and

Fig. 141.

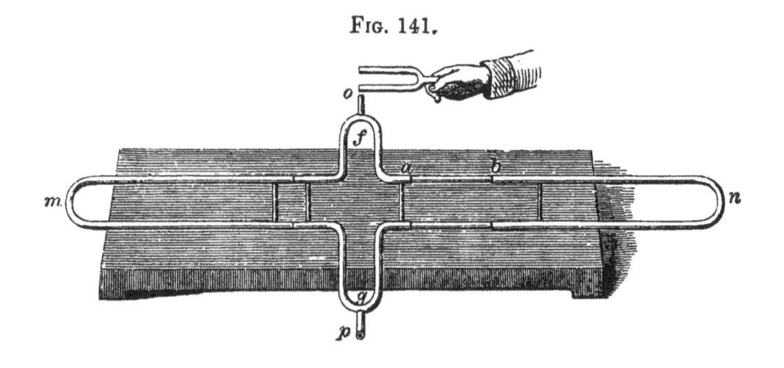

thus the sound-waves can be caused to pass over different distances in the two branches. Placing a vibrating tuning-fork at *o*, and the ear at *p*, when the two branches are of the same length, the waves through both reach the ear together, and the sound of the fork is heard. Drawing *a b* out, a point is at length attained where the sound of the fork is extinguished. This occurs when the distance *a b* is one-fourth of a wave-length; or in other words, when the whole right-hand branch is half a wave-length longer than the left-hand one. Drawing *b n* still further out, the sound is again heard; and when twice the distance *a b* amounts to a whole wave-length, it reaches a maximum. Thus according as the difference of both branches amounts to half a wave-length, or to a whole wave-length, we have interference or coincidence of the two series of sonorous waves. In practice the tube *o f* ought to be prolonged until the direct sound of the fork is unheard, the attention of the ear being then wholly concentrated on the sounds that reach it through the tube.

It is quite plain that the wave-length of any simple tone may be readily found by this instrument It is only necessary to ascertain the difference of path which produces complete interference. Twice this difference is the wave-length; and if the rate of vibration be at the same time known, we can immediately calculate the velocity of sound in air.

Each of the two forks now before you executes exactly 256 vibrations in a second, and when they are sounded together you have the perfect flow of unison. I now load one of them with a bit of wax, thus causing it to vibrate a little more slowly than its neighbour. Supposing, for the sake of simplicity, that the wax reduces the number of vibrations to 255 in a second, what must occur when the two forks are sounded together? If they start at the same moment, condensation coinciding with condensation, and rarefaction with rarefaction, it is quite manifest that this state of things cannot continue. The two forks soon begin to exert opposite actions upon the surrounding air. At the 128th vibration their phases are in complete opposition, one of them having gained half a vibration on the other. Here the one fork generates a condensation where the other generates a rarefaction; and the consequence is, that the two forks, at this particular point, completely neutralise each other. and we have no sound. From this point onwards, however, the forks support each other more and more, until, at the end of a second, when the one has completed its 255th, and the other its 256th vibration, the state of things is what it was at the commencement. Condensation then coincides with condensation, and rarefaction with rarefaction, the full effect of both sounds being produced upon the ear.

It is quite manifest, that under these circumstances we cannot have the continuous flow of perfect unison. We have, on the contrary, alternate reinforcements and diminutions of the sound. We obtain, in fact, the effect known to musicians by the name of *beats*, which, as here explained, are a result of interference.

I now load this fork still more heavily, by attaching a fourpenny-piece to the wax; the coincidences and interferences follow each other more rapidly than before; we have a quicker succession of beats. In our last experiment, the one fork accomplished one vibration more than the other in a second, and we had a single beat in the same time. In the present case, one fork vibrates 250 times, while the other vibrates 256 times in a second, and the number of beats per second is 6. A little reflection will make it plain, that in the interval required by the one fork to execute one vibration more than the other, a beat must occur; and inasmuch as, in the case now before us, there are six such intervals in a second, there must be six beats in the same time. In short, *the number of beats per second is always equal to the difference between the two rates of vibration.*

These beats may be produced by all sonorous bodies. The two tall organ-pipes now before you, when sounded together, give powerful beats. You notice that one of them is slightly longer than the other. Here are two other pipes, which are now in perfect unison, being exactly of the same length. But I have only to bring my finger near the embouchure of one of the pipes, fig. 142, to lower its rate of vibration, and produce these loud and rapid beats. The placing of my hand over the open top of one of the pipes also lowers its rate of vibration, and produces beats, which follow each other with augmented rapidity as I close the top of the pipe more and more. By a stronger blast I now bring out the two first harmonics of the pipes. These higher notes also interfere, and you have these sharper beats.

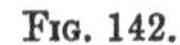

Fig. 142.

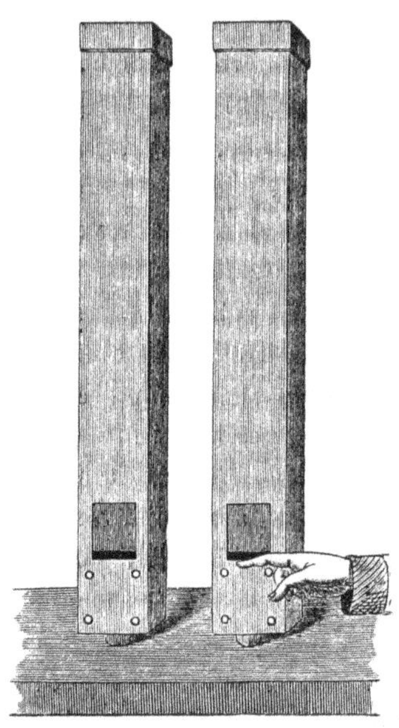

No more beautiful illustration of this phenomenon can be adduced than that furnished by two sounding flames. Two such flames are now before you, the tubes surrounding them being provided with telescopic sliders. There are at present no beats, because the tubes are not sufficiently near unison. I gradually lengthen the shorter tube by raising its slider. Rapid beats are now heard; now they are slower; now slower still; and now both flames sing together in perfect unison. Continuing the upward motion of the slider, I make the tube too long; the beats begin again, and quicken, until finally their sequence is so rapid as to appeal only as roughness to the ear. The flames, you observe, dance within their tubes in time to the beats. As already stated, these beats cause a silent flame within a tube to quiver when the voice is thrown to the proper pitch, and when the position of the flame is rightly chosen, the beats set it singing. With the flames of large rose-burners, and with tin tubes from 3 to 9 feet long, we obtain, as you perceive, beats of exceeding power.

You have just heard the beats produced by two organ-pipes nearly in unison with each other. Two similar pipes are now before you, fig. 143, each of which, however, is provided at its centre with a membrane intended to act upon a flame.* Two small tubes lead from the spaces closed by the membranes, and unite afterwards, the membranes of both the organ-pipes being in connection with

Fig. 143.

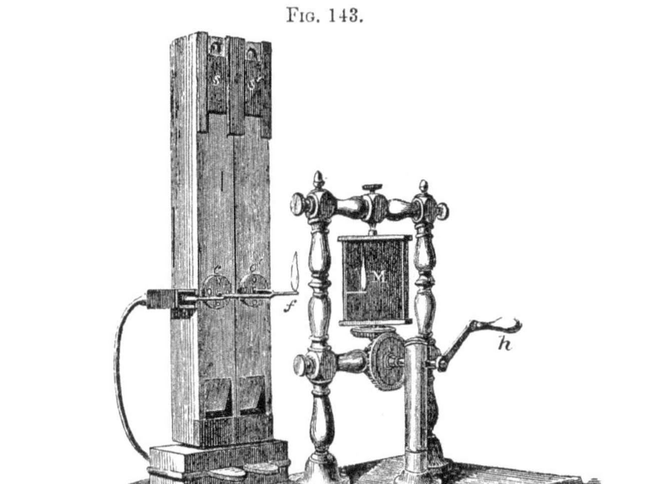

the same flame. By means of the sliders, *s*, *s*′, near the summits of the pipes, they are either brought into unison or thrown out of it at pleasure. I now sound both pipes. They are not in unison, and the beats they produce follow each other with great rapidity; while the flame connected with the central membranes dances in time to the beats. I bring them nearer to unison, the beats are now slow, and the flame at successive intervals

* Described in Lecture V.

withdraws its light and appears to exhale it. A process which reminds you of the inspiration and expiration of the breath is thus carried on by the flame. If the mirror now be turned, the flame produces a luminous band; continuous at certain places, but for the most part broken into distinct images of the flame. The continuous parts correspond to the intervals of interference, where the two sets of vibrations abolish each other.

Instead of permitting both pipes to act upon the same flame we may associate a flame with each of them. The deportment of the flames is then very instructive. Supposing both flames to be in the same vertical line, the one of them being exactly under the other. Bringing the pipes into unison, and turning the mirror, we resolve each flame into a chain of images, but we notice that the images of the one occupy the spaces between the images of the other. The periods of extinction of the one flame, therefore, correspond to the periods of rekindling of the other. The experiment proves, that when two unisonant pipes are placed thus close to each other, their vibrations are in opposite phases. The consequence of this is, that the two sets of vibrations permanently neutralise each other, so that at a little distance from the pipes you fail to hear the fundamental tone of either. For this reason we cannot, with any advantage, place close to each other several pipes of the same pitch in an organ.

In the case of beats, the amplitude of the oscillating air reaches a maximum and a minimum periodically. By the beautiful method of M. Lissajous we can illustrate optically this alternate augmentation and diminution of amplitude. Placing a large tuning-fork, T′, fig. 144, in front of the electric lamp, L, I receive upon the mirror of the fork a luminous beam, which is reflected back to the mirror of this second fork, T, and by it thrown on to the screen, where it forms a luminous disc. You notice that both forks are upright in this experiment. I draw my bow over the fork T; the beam, as in the experiments described in our second lecture, is tilted up and down, the disc upon the screen stretching to a luminous band three feet long. I now agitate the fork T; a moment's reflection will make it plain to you that

Fig. 144.

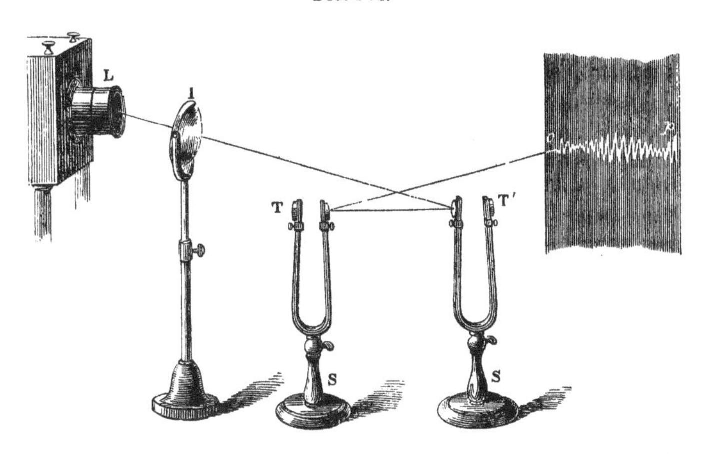

the one fork here may either act in concert with, or in opposition to the other, the band upon the screen being, in consequence, either lengthened or shortened. Whether it is so, or not, is a matter of accident. If, in drawing my bow over the second fork, I cause the vibrations of both to coincide in phase, I lengthen the band; if the phases are in opposition, total or partial neutralisation of one fork by the other will be the result. It so happens that the second fork here adds something to the action of the first, the band of light being now four feet long. I have tuned these forks as perfectly as possible. Each of them executes exactly 64 vibrations in a second; the initial relation of their phases remains, therefore, constant, and hence you notice a gradual shortening of the luminous band, exactly like that observed in the case of a single fork, during the subsidence of its vibration. The band at length dwindles to the original disc, which remains motionless upon the screen.

I now attach, with wax, a threepenny-piece to the prong of one of these forks, and thereby lower its rate of vibration. The phases of the two forks cannot now retain a constant relation to each other. One fork incessantly gains upon the other, and the consequence is that sometimes the phases of both will coincide, and at other times they will be in opposition. Observe the result. At the present moment the two forks conspire, and we have a luminous band four feet long upon the screen. This slowly contracts, drawing itself up to a mere disc; but the action halts here only during the moment of opposition. That passed, the forks begin again to assist each other, and the disc once more slowly stretches into a band. The action here is very slow. I quicken it by attaching a sixpence to the loaded fork. The band of light now stretches, and contracts in perfect rhythm. When the band is longest the two forks conspire; when it is shortest their phases are in opposition. The action, rendered here optically evident, is impressed upon the air of this room; its particles alternately vibrate and come to rest, and, as a consequence, beats are heard in perfect synchronism with the changes of the figure upon the screen.

The time which elapses from maximum to maximum, or from minimum to minimum, is that required for the one fork to perform one vibration more than the other. At present this time is about two seconds. In two seconds, therefore, one beat occurs. I augment the dissonance by increasing the load; the rhythmic lengthening and shortening of the band of light is now more rapid, while the intermittent hum of the forks is very audible. If you look at your watches, you will observe that there are now six elongations and shortenings in the interval taken up a moment ago by one; the beats are at the same time heard at the rate of three a second. By loading the fork still more, the alternations may be caused to succeed each other so rapidly that the lengthenings and shortenings can no longer be followed by the eye, while the beats, at the same time, cease to be individually distinct, and appeal as a kind of roughness to the ear.

In the experiments with a single tuning-fork, described in our second lecture, I received the beam reflected from the fork on a looking-glass, and, by turning the glass, caused the band of light upon the screen to stretch out into a long wavy line. I explained to you at the time that the loudness of the sound depended on the depth of those indentations. In the present instance we have not a continuous but an intermittent sound. If amplitude, therefore, expresses loudness, and if the image now before us be drawn out in a sinuous line, the sinuosities ought to be at some places deep, while at others they ought to vanish altogether. This is the case. By a little tact I cause the mirror of the fork T to turn through a small angle, and you have now before you a sinuous line composed of swellings and contractions, shown in part in fig. 144, but more fully in fig. 145, the

FIG. 145.

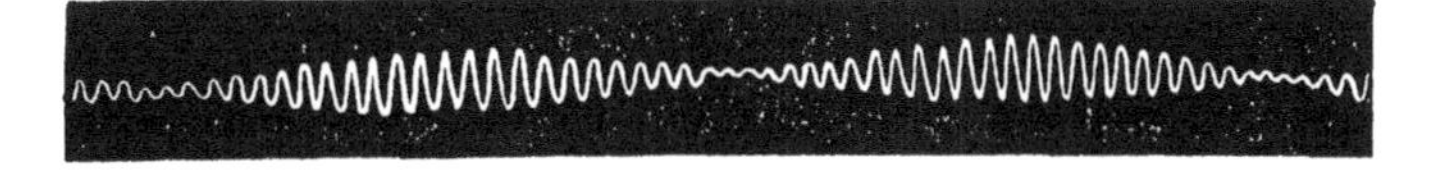

swellings corresponding to the periods of coincidence, and the contractions to the periods of interference.*

* The figure is but a meagre representative of the fact. The band of light was two inches wide, the depth of the sinuosities varying from three feet to zero.

Here, then, we have amply illustrated the general truth that two vibrating bodies, each of which separately produces a musical sound, can, when acting contemporaneously, abolish each other's action. It follows from this that when two vibrating bodies neutralise each other, we can, by quenching the vibrations of one of them, give sonorous effect to the other. It very often happens that when two tuning-forks, on their resonant cases, are vibrating in unison, the stoppage of one of them is accompanied by an augmentation of the sound. I can further illustrate this point by means of the vibrating bell exhibited in our fourth lecture. Placing its resonant tube in front of one of its nodes, you hear a sound, but nothing like what you hear when the tube is opposed to a ventral segment; the reason being that, the vibrations of a bell on the two sides of a nodal line are in opposite directions, and therefore interfere with each other. By introducing a glass plate between the bell and the tube I intercept the vibrations on one side of the nodal line; an instant augmentation of the sound is the consequence.

A disc is still better suited for this experiment. You have already learned that in the case of a vibrating disc every two adjacent sectors move at the same time in opposite directions. When the one rises the other falls, the nodal line marking the limit where there is neither rise nor fall. Hence, at the moment when any one of the sectors produces a condensation in the air above it, the adjacent sector produces a rarefaction in the same air. Interference, and a partial destruction of the sound of one sector by the other, is the result. And here let me introduce to your notice an instrument by which the late William Hopkins illustrated the principle of interference. The tube A B, fig. 146, divides at B into two branches. The end A of the tube is closed by a membrane. Scattering sand upon this membrane, and holding the ends of the branches over *adjacent* sectors of a vibrating disc, no motion (or at least an extremely feeble motion) of the sand is perceived. Here, in fact, the waves from both sectors neutralise each other, being generated by vibrations which are equal and opposite. Placing the ends of the two branches over *alternate* sectors of the disc, as in fig. 146, the sand is tossed from the membrane, proving that in this case we have coincidence of vibration on the part of the two sectors.

Fig. 146.

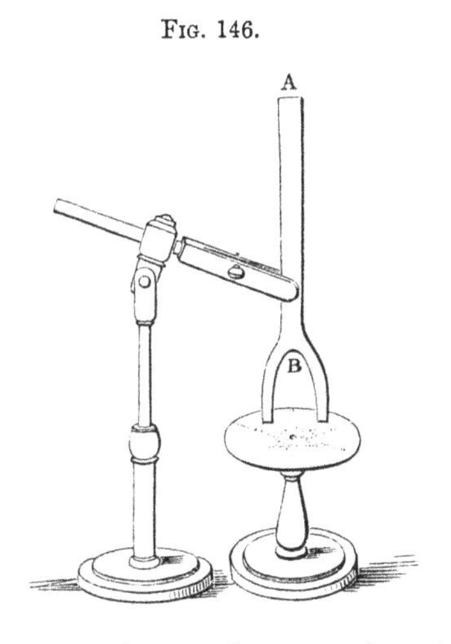

We are now prepared for a very instructive experiment which we owe to M. Lissajous. I divide this brass disc into six vibrating sectors; and bringing the palm of my hand near any one of the sectors I intercept its vibrations. The sound is augmented. Placing my two hands over two *adjacent* sectors, you notice no increase of the sound. Placing them, however, over *alternate* sectors, as in fig. 147, a striking augmentation of the sound is the consequence. By simply lowering and raising my two hands, I produce these marked variations of intensity. By the approach

Fig. 147.

of my hands I intercept the vibrations of the two sectors; their interference right and left being thus abolished, the remaining sectors sound more loudly. Passing my single hand to and fro along the surface, you also hear a rise and fall of the sound. It rises when my hand is over a vibrating sector; it falls when the hand is over a nodal line. Thus, by sacrificing a portion of the vibrations, we make the residue more effectual. Experiments similar to these may be made with light and radiant heat. If of two beams of the former, which destroy each other by interference, one be removed, light takes the place of darkness; and if of two interfering beams of the latter, one be intercepted, heat takes the place of cold.

You must have remarked the almost total absence of sound on the part of a vibrating tuning-fork when held free in the hand. The feebleness of the fork as a sounding body arises in great part from interference. The prongs always vibrate in opposite directions, one producing a condensation where the other produces a rarefaction, a destruction of sound being the consequence. By simply passing a pasteboard tube over one of the prongs of the fork, its vibrations are in part intercepted, and an augmentation of the sound is the result. The single prong is thus proved to be more effectual than the two prongs. There are positions in which the destruction of the sound of one prong by that of the other is *total*. These positions are easily found by striking the fork, and turning it round before the ear. When the back of the prong is parallel to the ear the sound is heard; when the side surfaces of both prongs are parallel to the ear the sound is also heard; but when the corner of a prong is carefully presented to the ear the sound is utterly destroyed. During one complete rotation of the fork we find four positions where the sound is thus obliterated.

Let s s' represent the two ends of the tuning-fork, looked down upon as it stands upright. When the ear is placed at *a* or *b*, or at *c* or *d*, the sound is heard. Along the four dotted lines, on the contrary, the waves generated by the two prongs completely neutralise each other, and along these lines nothing is heard. These lines have been proved by Weber to be hyperbolic curves; and this must be their character according to the principle of interference.

FIG. 148.

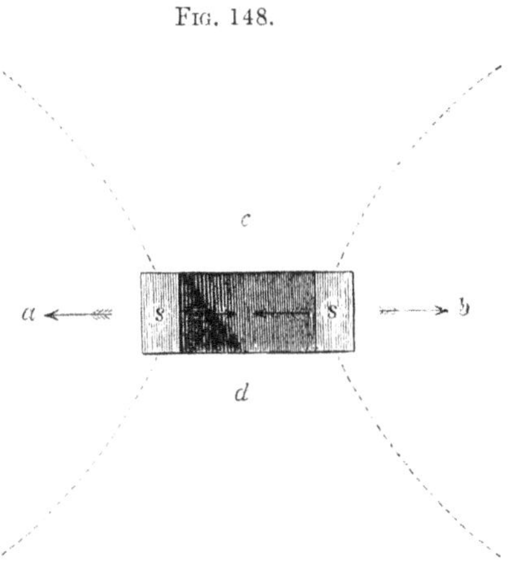

This remarkable case of interference, which was first noticed by Dr. Thomas Young, and thoroughly investigated by the brothers Weber, may be rendered audible to you all by means of resonance. I have here a jar which powerfully resounds to this tuning-fork. Bringing the fork over the jar, I cause it to rotate slowly. In four positions you have this loud resonance; in four others absolute silence, alternate risings and fallings of the sound accompanying the fork's rotation. While the fork is over the jar with

FIG. 149.

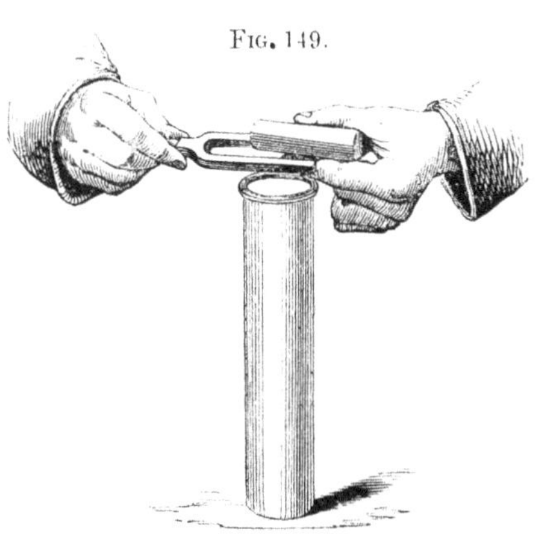

its corner downwards, and the sound entirely extinguished, I pass a pasteboard tube over one of its prongs, as in fig. 149; a loud resonance announces the withdrawal of the vibrations of that prong. To obtain this effect the fork must be held over the centre of the jar, so that the air shall be symmetrically distributed on both sides of it. Moving the fork from the centre towards one of the sides, without altering its inclination in the least, you obtain a forcible sound. Interference, however, is also possible near the side of the jar. Holding the fork, not with its corner downwards, but with both its prongs in the same horizontal plane, a position is soon found near the side of the jar where the sound is extinguished. In passing completely from side to side over the mouth of the jar, two such places of interference are discoverable.

A variety of experiments will suggest themselves to the reflecting mind, by which the effect of interference may be illustrated. It is easy, for example, to find a jar which resounds to a vibrating plate. I place such a jar over a vibrating segment of this disc, the resonance is powerful; placed over a nodal line, the resonance is entirely absent. I interpose a piece of pasteboard between the jar and disc, so as to cut off the vibrations on one side of the nodal line; the jar instantly resounds to the vibrations of the other. Again, holding two forks, which vibrate with the same rapidity, over two resonant jars, the sound of both flows forth in unison. When a bit of wax is attached to one of the forks, powerful beats are heard. Removing the wax, the unison is restored. I place one of these unisonant forks in the flame of a spirit lamp and warm it; its elasticity is changed by its change of temperature, and now it produces these long loud beats with its unwarmed fellow. I load the cold fork a little and find the unison restored; the heat, therefore, has diminished the elasticity of the steel.* I

* In his admirable experiments on tuning, Scheibler found in the beats a test of differences of temperature of exceeding delicacy.

now cool the fork, restore it to its resonant case, and draw my bow across it. Both the wood and the air of the case are thrown into resonant vibration. In fact, the case is so constructed that the air within it resounds to the fork. I load the other fork, and bring it near the mouth of the case, as in fig. 150; loud beats are the consequence. I divide this jar by a vertical diaphragm, and bring one of the forks over one of its halves, and the

Fig. 150.

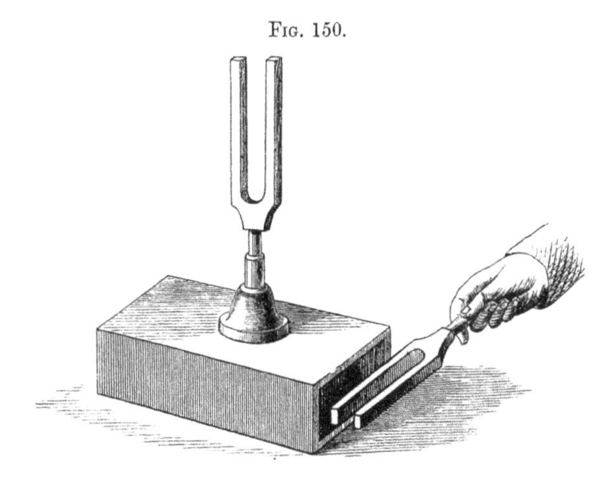

other fork over the other. The two semi-cylinders of air produce beats by their interference. I now remove the diaphragm; the beats continue as loud as before, one half of the same column of air interfering with the other.*

The intermittent sounds of certain bells, heard more especially when their tones are subsiding, is an effect of interference. The bell, through lack of symmetry, as explained in our fourth lecture, vibrates in one direction a little more rapidly than in the other, and beats are the consequence of the coalescence of the two different rates of vibration.

* Mr. Wheatstone and Sir John Herschel, I believe, made this experiment independently.

RESULTANT TONES.

We have now to turn from this question of interference to the consideration of a new class of musical sounds, of which the beats were long considered to be the progenitors. The sounds here referred to require for their production the union of two distinct musical tones. Where such union is effected, under the proper conditions, *resultant tones* are generated, which are quite distinct from the primaries concerned in their production. These resultant tones were discovered in 1745, by a German organist named Sorge, but the publication of his discovery attracted little attention. They were discovered independently in 1754 by the celebrated Italian violinist Tartini, and after him have been called Tartini's tones.

To produce these tones it is desirable, if not necessary, that the two primary tones shall be of considerable intensity. Helmholtz prefers the syren to all other means of exciting them, and with this instrument they are very readily obtained. It requires some attention at first, on the part of the listener, to single out the resultant tone from the general mass of sound; but with a little practice, this is readily accomplished; and though the unpractised ear may fail, in the first instance, thus to analyse the sound, the clang-tint is influenced in an unmistakable manner by the admixture of resultant tones. Before me is Dove's syren; I set it in rotation, and open two series of holes at the same time. With all the attention I can bestow, I am as yet unable to hear the least symptom of a resultant tone. I urge the instrument to greater rapidity, and now for the first time I hear a dull low droning mingling with the two primary sounds. I urge the syren to greater speed, and as I do so, the low resultant tone rises rapidly in pitch, and now to me, who stand close to the instrument, it is very audible. The two series of holes here open, number 8 and 12 respectively. The resultant tone is in this case precisely such as we should obtain if a series of 4 holes were opened in this rotating disc; that is to say, it is an octave below the deepest of the two primaries. I now open two other series of orifices, numbering 12 and 16 respectively. The resultant tone is here quite audible, and it also would be given by a series of 4 holes in the rotating disc. Its rate of vibration is therefore one-third of the rate of the deepest of the two primaries. Again I open two series, numbering respectively 10 and 16. The resultant of these two tones would be given by a series of 6 orifices in our rotating disc. In all these cases, *the resultant tone is that which corresponds to a rate of vibration equal to the difference of the rates of the two primaries.*

When I speak here of the resultant tone, I mean the one actually heard in the experiment. But this is not the only resultant. With finer methods of experiment other resultant tones are proved to exist. Those on which we have now fixed our attention are, however, the most important, on account of their intensity. They are called *difference tones* by Helmholtz, in consequence of the law above mentioned.

To bring these resultant tones audibly forth, the primaries must, as already stated, be forcible. When they are feeble the resultants are unheard. I am acquainted with no method of exciting these tones more simple and effectual than a pair of suitable singing flames. Before me are two flames of common gas, over which I place two glass tubes, provided with paper sliders, by which their lengths may be varied within certain limits. Both flames are now emitting powerful notes,—self-created, self-sustained, and requiring no muscular effort on the part of the observer to keep them going. The length of the shorter tube at this moment is $10\frac{3}{8}$ inches, that of the other is 11·4 inches. I hearken to the sound, and in the midst of the shrillness detect a very deep resultant tone. The reason of its depth is manifest: the two tubes being so nearly alike in length, the difference between their vibrations is small, and the note corresponding to this difference, therefore, low in pitch. But I now lengthen one of the tubes by means of its slider; the resultant tone rises gradually, and now it swells to an extent sufficient to render it audible to many of you. I shorten the tube; again the resultant tone falls, and thus by alternately raising and lowering the slider, I cause the resultant tone, to rise and sink in accordance with the law which makes the number of its vibrations the difference between the numbers of its two primaries.

We can determine, with ease, the actual number of vibrations corresponding to any one of those resultant tones. The sound of the flame is that of the open tube which surrounds it, and we know that the length of such a tube is half that of the sonorous wave that it produces. The wave-length, therefore, corresponding to our $10\frac{3}{8}$-inch tube is $20\frac{3}{4}$ inches. The velocity of sound in air of the temperature of this room is 1,120 feet a second. Bringing these feet to inches, and dividing by $20\frac{3}{4}$, we find the number of vibrations corresponding to a length of $10\frac{3}{8}$ inches to be 648 per second.

But it must not be forgotten here, that the air in which the vibrations are actually executed is much more elastic than the air of this room. The flame heats the air of the tube, and the vibrations must, therefore, be executed more rapidly than they would be in an ordinary organ-pipe of the same length. To determine the actual number of vibrations, we must fall back upon our syren; and with this instrument I find that the air within our $10\frac{3}{8}$-inch tube executes 717 vibrations in a second. The difference of 69 vibrations a second is due to the heating of the aërial column. Carbonic acid and aqueous vapour are, moreover, the products of the flame's combustion, and their presence must also affect the rapidity of the vibration.

Determining in the same way the rate of vibration of our 11·4-inch tube, we find it to be 667 per second; the difference between this number and 717 is 50, which expresses the rate of vibration corresponding to our first deep resultant tone.

But this number does not mark the limit of audibility. Permitting the 11·4-inch tube to remain as before, I now lengthen its neighbour until the resultant tone sinks near the limit of my power of hearing. I will not push matters beyond the point of certainty, and now, when my shorter tube measures 11 inches, I can still plainly hear the deep sound of the resultant tone. The number of vibrations per second executed in this 11-inch tube I find to be 700. We have already found the number executed in the 11·4-inch tube to be 667; hence $700 - 667 = 33$, which is the number of vibrations corresponding to the resultant tone now plainly heard when I converge my attention upon it. We here come very near the limit which Helmholtz has fixed as that of musical audibility. Again, I take this tube, $17\frac{3}{8}$ inches in length, and cause its sound to combine with that of the $10\frac{3}{8}$-inch tube. A resultant tone of higher pitch than any previously heard is the consequence. Now the actual number of vibrations executed in this longer tube is 459; and we have already found the vibrations of our $10\frac{3}{8}$-inch tube to be 717; hence $717 - 459 = 258$, which is the number corresponding to the resultant tone now audible. This note is almost exactly that of one of our tuning-forks, which, you will remember, vibrates 256 times in a second.

And now let us avail ourselves of a beautiful check

which this result suggests to us. Here is the well-known fork which vibrates at the rate just mentioned. It is mounted on its case, and I touch it with my bow so lightly that the sound alone could hardly be heard; but it instantly coalesces with the resultant tone, and the beats produced by their combination are clearly audible. By loading my fork, and thus altering its pitch, or by drawing up my paper slider, and thus altering the pitch of the flame, I can alter the rate of these beats, exactly as you have seen me alter it when comparing two primary tones together. By slightly varying the size of the flame the same effect is produced. You cannot fail to observe how beautifully these results play into each other.

Standing midway between the syren and a shrill singing flame, and gradually raising the pitch of the syren, the resultant tone soon makes itself heard, sometimes swelling out with extraordinary power. When a pitch-pipe is blown near the flame, the resultant tone is also heard, seeming, in this case, to originate in the ear itself, or rather in the brain. By gradually drawing out the stopper of the pipe, the pitch of the resultant tone is caused to vary in accordance with the law already enunciated.

The resultant tones produced by the combination of the ordinary harmonic intervals * are given in the following table:—

Interval	Ratio of vibrations	Difference	The resultant tone is deeper than the lowest primary tone by
Octave	1 : 2	1	0
Fifth	2 : 3	1	an octave
Fourth	3 : 4	1	a twelfth
Major third . .	4 : 5	1	two octaves
Minor third . .	5 : 6	1	two octaves and a major third
Major sixth . .	3 : 5	2	a fifth
Minor sixth . .	5 : 8	3	major sixth

* A subject to be dealt with in our next lecture.

The celebrated Thomas Young thought that these resultant tones were due to the coalescence of rapid beats, which linked themselves together like the periodic impulses of an ordinary musical note. This explanation harmonised with the fact that the number of the beats, like that of the vibrations of the resultant tone, is equal to the difference between the two sets of vibrations which produce the beats. This explanation, however, is insufficient. The beats tell more forcibly upon the ear than any continuous sound. They can be plainly heard when each of the two sounds that produce them has ceased to be audible. This depends in part upon the sense of hearing, but it also depends upon the fact that when two notes of the same intensity produce beats, the amplitude of the vibrating air-particles is at times destroyed, and at times doubled. But by doubling the amplitude we quadruple the intensity of the sound. Hence when two notes of the same intensity produce beats, *the sound incessantly varies between silence and a tone of four times the intensity of either of the interfering ones.*

If therefore the resultant tones were due to the beats of their primaries, they ought to be heard, even when the primaries are feeble. But they are not heard, under these circumstances. This fact led Helmholtz to reinvestigate the subject. I have already had occasion to state to you, that when several sounds traverse the same air, each particular sound passes through the air as if it alone were present, that amid composite sounds, each particular element of the composition asserts its own individuality, and nothing more. Now, this is only in strictness true when the amplitudes of the oscillating particles are infinitely small. Guided by pure reasoning, the mathematician arrives at this result. The law is also practically true when the disturbances are *extremely* small; but it is *not* true after they have passed a certain limit. Vibrations which produce a large amount of disturbance give birth to secondary waves, which appeal to the ear as resultant tones. Having proved this, Helmholtz inferred further that there are also resultant tones formed by the sum of the primaries, as well as by their difference. He thus discovered his *summation tones* before he had heard them; and bringing his result to the test of experiment, he found that these summation tones have a real physical existence. They are not to be explained by Young's theory, but they receive a complete explanation by that of Helmholtz.

Another consequence of this departure from the law of superposition is, that a single sounding body, which disturbs the air beyond the limits of the law of the superposition of vibrations, also produces secondary waves, which correspond to the harmonic tones of the vibrating body. For example, the rate of vibration of the first overtone of a tuning-fork, as stated in our fourth lecture, is $6\frac{1}{4}$ times the rate of the fundamental tone. But Helmholtz distinctly proves that a tuning-fork, not excited by a bow, but vigorously struck against a pad, emits the *octave* of its fundamental note, this octave being due to the secondary waves set up when the limits of the law of superposition have been exceeded.

These considerations make it probably evident to you that a coalescence of musical sounds is a far more complicated dynamical condition than you have hitherto supposed it to be. In the music of an orchestra, not only have we the fundamental tones of every pipe and of every string, but we have the overtones of each, sometimes audible as far as the sixteenth in the series. We have also resultant tones; both difference tones and summation tones; all trembling through the same air, all knocking at the self-same tympanic membrane. We have fundamental tone interfering with fundamental tone; we have overtone interfering with overtone; we have resultant tone interfering with resultant tone. And besides this, we have the members of each class interfering with the members of every other class. The imagination retires baffled from any attempt to realise the physical condition of the atmosphere through which those sounds are passing. And, as we shall learn in our next lecture, the aim of music, through the centuries during which it has ministered to the pleasure of man, has been to arrange matters empirically, so that the ear shall not suffer from the discordance produced by this multitudinous interference. The musicians engaged in this work knew nothing of the physical facts and principles involved in their efforts; they knew no more about it than the inventors of gunpowder knew about the law of atomic proportions. They tried and tried till they obtained a satisfactory result, and now, when the scientific mind is brought to bear upon the subject, order is seen rising through the confusion, and the results of pure empiricism are found to be in harmony with natural law.

SUMMARY OF LECTURE VII.

When several systems of waves proceeding from distinct centres of disturbance pass through water or air, the motion of every particle is the algebraic sum of the several motions impressed upon it.

In the case of water, when the crests of one system of waves coincide with the crests of another system: higher waves will be the result of the coalescence of the two systems. But when the crests of one system coincide with the sinuses, or furrows, of the other system, the two systems, in whole or in part, destroy each other.

This mutual destruction of two systems of waves is called *interference.*

The same remarks apply to sonorous waves. If in two systems of sonorous waves condensation coincides with condensation, and rarefaction with rarefaction, the sound produced by such coincidence is louder than that produced by either system taken singly. But if the condensations of the one system coincide with the rarefactions of the other, a destruction, total or partial, of both systems is the consequence.

Thus, when two organ-pipes of the same pitch are placed near each other on the same wind-chest and thrown into vibration, they so influence each other, that as the air enters the embouchure of the one it quits that of the other. At the moment, therefore, the one pipe produces a condensation the other produces a rarefaction. The sounds of two such pipes mutually destroy each other.

When two musical sounds of nearly the same pitch are sounded together the flow of the sound is disturbed by *beats.*

These beats are due to the alternate coincidence and interference of the two systems of sonorous waves. If the two sounds be of the same intensity their coincidence produces a sound of four times the intensity of either; while their interference produces absolute silence.

The effect, then, of two such sounds, in combination, is a series of shocks, which we have called 'beats,' separated from each other by a series of 'pauses.'

The rate at which the beats succeed each other is equal to the difference between the two rates of vibration.

When a bell or disc sounds, the vibrations on opposite sides of the same nodal line partially neutralise each other; when a tuning-fork sounds the vibrations of its two prongs in part neutralise each other. By cutting off a portion of the vibrations in these cases the sound may be intensified.

When a luminous beam, reflected on to a screen from two tuning-forks producing beats, is acted upon by the vibrations of both, the intermittence of the sound is announced by the alternate lengthening and shortening of the band of light upon the screen.

The law of the superposition of vibrations above enunciated is strictly true only when the amplitudes are exceedingly small. When the disturbance of the air by a sounding body is so violent that the law no longer holds good, secondary waves are formed which correspond to the harmonic tones of the sounding body.

When two tones are rendered so intense as to exceed the limits of the law of superposition, their secondary waves combine to produce *resultant tones.*

Resultant tones are of two kinds; the one class corresponding to rates of vibration equal to the difference of the rates of the two primaries; the other class corresponding to rates of vibration equal to the sum of the two primaries. The former are called *difference tones,* the latter *summation tones.*

LECTURE VIII.

COMBINATION OF MUSICAL SOUNDS—THE SMALLER THE TWO NUMBERS WHICH EXPRESS THE RATIO OF THEIR RATES OF VIBRATION, THE MORE PERFECT IS THE HARMONY OF TWO SOUNDS—NOTIONS OF THE PYTHAGOREANS REGARDING MUSICAL CONSONANCE—EULER'S THEORY OF CONSONANCE—PHYSICAL ANALYSIS OF THE QUESTION—THEORY OF HELMHOLTZ—DISSONANCE DUE TO BEATS—INTERFERENCE OF PRIMARY TONES AND OF OVERTONES—GRAPHIC REPRESENTATION OF CONSONANCE AND DISSONANCE—MUSICAL CHORDS—THE DIATONIC SCALE—OPTICAL ILLUSTRATION OF MUSICAL INTERVALS—LISSAJOUS' FIGURES—SYMPATHETIC VIBRATIONS—MECHANISM OF HEARING—SCHULTZE'S BRISTLES—THE OTOLITES—CORTI'S FIBRES—CONCLUSION.

THE subject of this day's lecture has two sides, the one physical, the other æsthetical. We have this day to study the question of musical consonance—to examine musical sounds in definite combination with each other; and to unfold the reason why some combinations are pleasant and others unpleasant to the ear.

Here are two tuning-forks mounted on their resonant cases. I draw a fiddle-bow across them in succession: they are now sounding together, and their united notes reach your ears as the note of a single fork. Each of these forks executes 256 vibrations in a second. Two musical sounds flow thus together in a perfectly blended stream, and produce this *perfect unison* when the ratio of their vibrations is as 1 : 1.

Here are two other forks, which I cause to sound by the passage of the bow. These two notes also blend sweetly and harmoniously together. By means of our syren I have already determined the rates of vibration of the forks, and found that this large one executes 256 vibrations a second, while the small one executes 512. For every single wave, therefore, sent to the ear by the one fork two waves are sent by the other. I need not tell the musicians present that the combination of sounds now heard is that of a fundamental note and its octave; nor need I tell them that, next to the perfect unison, these two notes, the ratio of whose vibrations is as 1 : 2, blend most harmoniously together.

I now throw another pair of forks into contemporaneous vibration. The combination of the two sounds is very pleasing to the ear, but the consonance is hardly so perfect as in the last instance. There is a barely perceptible roughness here, which is absent when a note and its octave are sounded. The roughness, however, is too insignificant to render the combination of the two notes anything but agreeable. I look to the numbers stamped upon these two forks, and find that one of them executes 256 and the other 384 vibrations in a second. These two numbers are to each other in the ratio of 2 : 3; one of the forks, therefore, sends two waves and the other three waves to the ear in the same interval of time. The musicians present know that the two notes now sounding are separated from each other by the musical interval called a fifth. Next to the octave this is the most pleasing combination.

I once more change the forks, and sound two others simultaneously. The combination is still agreeable, but not so agreeable as the last. The roughness there incipient is here a little more pronounced. One of these forks executes 384 vibrations, the other 512 vibrations, in a second; the two numbers standing to each other in the ratio of 3 : 4. This interval the musicians present recognise as a fourth. Thus, then, with perfect unison the ratio of the vibrations is as 1 : 1; with a note and its octave it is 1 : 2; with a note and its fifth it is 2 : 3; and with a note and its fourth it is 3 : 4. We here notice the gradual development of the remarkable law that *the combination of two notes is the more pleasing to the ear, the smaller the two numbers which express the ratio of their vibrations.* I pass on to two forks whose rates of vibration are in the ratio of 4 : 5, or a major third apart; the harmony is less perfect than in any of the cases which we have examined. With a ratio of 5 : 6, or that of a minor third, it is usually less perfect still, and we now approach a limit beyond which a musical ear will not tolerate the combination of two sounds. Here, for example, are two forks whose vibrations are in the ratio of 13 : 14; you would pronounce their combination altogether discordant. It must not be imagined that the choice of these combinations was determined by scientific knowledge. They were chosen empirically, and in consequence of the pleasure which they gave, long before anything was known regarding their numerical simplicity.

Pythagoras made the first step towards the physical explanation of these musical intervals. This great philosopher stretched a string, and then divided it into three equal parts. At one of its points of division he fixed it firmly, thus converting it into two, one of which was twice the length of the other. He sounded the two sections of the string simultaneously, and found the note emitted by the short section to be the higher octave of that emitted by the long one. He then divided his string into two parts, bearing to each other the proportion of 2 : 3, and found that the notes were separated by an interval of a fifth. Thus, dividing his string at different points, Pythagoras found the so-called consonant intervals in music to correspond with certain lengths of his string; and he made the extremely important discovery, that the simpler the ratio of the two parts into which the string was divided, the more perfect was the harmony of the two sounds. Pythagoras went no further than this, and it remained for the investigators of a subsequent age to show that the strings act in this way in virtue of the relation of their lengths to the number of their vibrations. Why simplicity should give pleasure remained long an enigma, the only pretence of a solution being that of Euler, which, briefly expressed, is, that the human soul takes a constitutional delight in simple calculations.

The double syren, the figure of which I reproduce here, fig. 151, enables us to obtain all these combinations and a great many others. And this instrument possesses over all others the advantage that, by simply counting the number of orifices corresponding respectively to any two notes, we obtain immediately the ratio of their rates of vibration. We need not, therefore, rely upon results obtained with tuning-forks which were stamped in Paris: we have, in our own hands, the means of arriving at absolute certainty regarding the combinations of musical sounds. But before proceeding to these combinations I will enter a little more fully into the action of the double syren than I have hitherto deemed necessary or desirable.

The instrument, as you know, consists of two of Dove's syrens, C′ and C, connected by a common axis, the upper syren C′ being turned upside down. Each syren is provided with four series of apertures, numbering as follows :—

	Upper syren. Number of apertures.	Lower syren. Number of apertures.
1st Series . . .	16	18
2nd Series . . .	15	12
3rd Series . . .	12	10
4th Series . . .	9	8

The number 12, it will be observed, is common to both syrens. I now open the two series of 12 orifices each, and urge air through the instrument; both sounds flow together in perfect unison; the unison being maintained however highly the pitch may be exalted. You have, however, learned in our second lecture that by turning the handle of the upper syren, we can cause the orifices in its wind-chest C′ either to meet those of its rotating

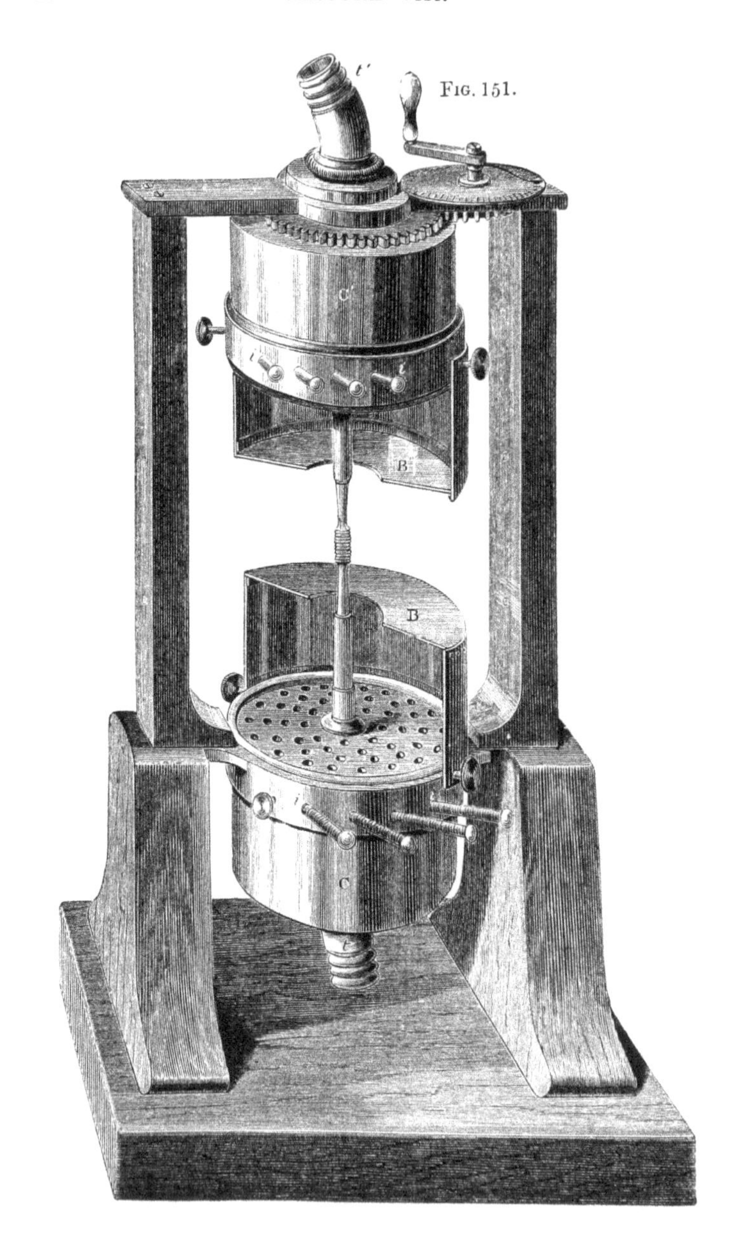

Fig. 151.

disc, or to retreat from them, and that this places it in our power either to raise or to lower, within small limits, the pitch of the upper syren. This change of pitch instantly announces itself by beats. The more rapidly the handle is turned, the more is the tone of the upper syren raised above or depressed below that of the lower one, and, as a consequence, the more rapid are the beats. Now the rotation of the handle is so related to the rotation of the wind-chest c′, that when the handle turns through half a right angle the wind-chest turns through $\frac{1}{6}$th of a right angle, which is equal to $\frac{1}{24}$th of its whole circumference. But in the case now before us, where the circle is perforated by 12 orifices, the rotation through $\frac{1}{24}$th of its circumference causes the apertures of the upper wind-chest to be closed at the precise moments when those of the lower syren are opened, and *vice versâ*. It is plain, therefore, that the intervals between the puffs of the lower syren, which correspond to the rarefactions of its sonorous waves, are here filled by the puffs, or condensations, of the upper syren. In fact, the condensations of the one coincide here with the rarefactions of the other, and the absolute extinction of the sounds of both syrens is the consequence.

I may seem to you to have exceeded the truth here, for when the handle is placed in the position which corresponds to absolute extinction, you still hear a sound. And when I turn the handle thus continuously, though alternate swellings and sinkings of the tone occur, the sinkings by no means amount to absolute silence. The reason is this. The sound of this syren is a highly composite one. By the suddenness and violence of its shocks, not only does it produce waves corresponding to the number of its orifices, but the aërial disturbance breaks up into secondary waves, which associate themselves with the primary waves of the instrument, exactly as the harmonics of a string, or of an open organ-pipe, mix with their fundamental tone. When the syren sounds, therefore, it emits, besides the fundamental tone, its octave, its twelfth, its double octave, and so on. That is to say, it breaks the air up into vibrations which have twice, three times, four times, &c., the rapidity of the fundamental one. Now, by turning the upper syren through $\frac{1}{24}$th of its circumference, we extinguish utterly the fundamental tone. But we do not extinguish its octave.* Hence, when I turn the handle into the position which corresponds to the extinction of the fundamental, instead of obtaining silence, I obtain the full first harmonic of the instrument.

You will observe that Helmholtz has surrounded both his upper and his lower syren with circular brass boxes, B, B′, each composed of two halves, one of which is removed in each of the figures. These boxes exalt by their resonance the fundamental tone of the instrument, and enable us to follow its variations much more easily than if it were not thus exalted. It requires a certain rapidity of rotation to reach the maximum resonance of the brass boxes, but when this speed is attained, the fundamental tone swells out with greatly augmented force, and, if the handle be then turned, the beats succeed each other with extraordinary power.

Still, as I have said, the pauses between the beats of the fundamental tone are not intervals of absolute silence, but are filled by the higher octave; and this renders caution necessary when the instrument is employed to determine rates of vibration. Wishing to determine the number of times a small singing flame extinguished and relighted itself in a second, I once placed a syren at some distance from the flame, sounded the instrument, and after a little time observed the flame dancing in synchronism with audible beats. I took it for granted

* Nor indeed any of those tones whose rates of vibration are *even* multiples of the rate of the fundamental.

that unison was nearly attained, and, under this assumption, determined the rate of vibration. The number obtained was surprisingly low; indeed I had reason to know that it could not be more than half what it ought to be. What was the reason? Simply this; I was dealing, not with the fundamental tone of the syren but with its higher octave. This octave and the flame produced beats by their coalescence, and hence the counter of the instrument, which recorded the rate, not of the octave, but of the fundamental, gave a number which was only half the true one. I afterwards raised the fundamental tone itself to unison with the flame. On approaching unison beats were again heard, and the jumping of the flame proceeded with an energy greater than that observed in the case of the octave. The counter of the instrument then recorded the accurate rate of the flame's vibration.

In fact, the tones first heard in the case of the syren are always overtones. They attain sonorous continuity sooner than the fundamental, flowing as smooth musical sounds while the fundamental tone is still in a state of intermittence. The instrument is, however, so delicately constructed that a rate of rotation is almost immediately attained which raises the fundamental tone above its fellows; and if you seek by making your blast feeble to keep the speed of rotation low, it is at the expense of the intensity of the overtones. Hence the desirability, if we wish to examine the overtones, of devising some means by which a strong blast and slow rotation shall be possible.

Helmholtz caused a spring to press lightly against the disc of the syren, and thus raising by slow degrees the speed of rotation, he was able to notice the predominance of the overtones at the commencement, and the final triumph of the fundamental tone. He did not trust to the direct observation of pitch, but he determined the tone by the number of beats corresponding to one revolution of the handle of the upper syren. Supposing 12 orifices to be opened above and 12 below, the motion of the handle through 45° produces interference and extinguishes the fundamental tone. The *coincidences* of that tone occur at the end of every rotation of 90°. Hence, for the fundamental tone there must be *four* beats for every 360°, or for every complete revolution of the handle. Now Helmholtz, when he made the arrangement just described, found that the first beats heard numbered, not 4 but 12 for every revolution. They were, in fact, the beats of the second overtone whose rate of vibration is three times that of the fundamental. These beats continued as long as the number of impulses per second did not exceed 30 or 40. Within this interval the second overtone was comparatively so powerful, that its beats were heard to the practical exclusion of the others. Between 40 and 80, the beats fell from 12 to 8 for every revolution of the handle. Within this interval the first overtone, or the octave of the fundamental tone, was the most powerful, and made the beats its own. Not until the impulses exceeded 80 per second did the beats sink to 4 per revolution. In other words, not until the speed of rotation had passed this limit, was the fundamental tone able to assert its superiority over its companions.

The instrument before you is now so arranged, that with a proper speed of rotation the fundamental tone is smooth and powerful. And now we will combine our tones in definite order, while the cultivated ears here present shall judge of their musical relationship. You have already heard the flow of perfect unison when the two series of 12 orifices each were opened. I now open a series of 8 holes in the upper, and a series of 16 in the lower syren. The musical ear, as already remarked, fixes the interval as an octave. If I open a series of 9 holes in the upper and of

18 holes in the lower syren, the interval is still an octave. This proves that the interval is not disturbed by altering the *absolute rates* of vibration so long as the *ratio* of the two rates remains the same. The same truth is more strikingly illustrated by commencing with a low speed of rotation, and urging the syren to its highest pitch; as long as the orifices are in the ratio of 1 : 2, we retain the constant interval of an octave. Again, I open a series of 10 holes in the upper and of 15 in the lower syren. The ratio here is as 2 : 3, and every musician present knows that this is the interval of a fifth. If I open a series of 12 holes in the upper and a series of 18 in the lower syren I do not change the interval. Opening two series of 9 and 12, or of 12 and 16, we obtain an interval of a fourth ; the ratio in both these cases being as 3 : 4. In like manner these two series of 8 and 10, or these other two of 12 and 15, give us the interval of a major third ; the ratio in this case being as 4 : 5. If, moreover, I open two series of 10 and 12, or of 15 and 18, I obtain the interval of a minor third, this interval corresponding to the ratio 5 : 6.

Now these experiments amply illustrate two things :— firstly, that a musical interval is determined not by the absolute number of vibrations of the two combining notes, but by the ratio of their vibrations. Secondly,—and here I can confidently appeal to the musical ears that have listened to these combinations —that the smaller the two numbers which express the ratio of the two rates of vibration, the more perfect is the consonance of the two sounds. The most perfect consonance is the unison 1 : 1 ; next comes the octave 1 : 2; after that the fifth 2 : 3 ; then the fourth 3 : 4; then the major third 4 : 5 ; and finally the minor third 5 : 6. It is also in my power to open two series numbering, respectively, 8 and 9 orifices: this interval corresponds to *a tone* in music. It is a dissonant combination. I can also open two series which number respectively 15 and 16 orifices; this is the interval of a *semi-tone*: it is a very sharp and grating dissonance.

Whence then does this arise? Why should the smaller ratio express the more perfect consonance? The attempts to answer this question were of two kinds—metaphysical and physical. The Pythagoreans found intellectual repose in the answer 'all is number and harmony.' The numerical relations of the seven notes of the musical scale were also thought by them to express the distances of the planets from their central fire; hence the choral dance of the worlds, the 'music of the spheres,' which, according to his followers, Pythagoras was the only mortal privileged to hear. And might we not in passing contrast this glorious superstition with those which have taken hold of the human fantasy in our day? Were the character which superstition assumes in different ages, an indication of man's advance or retrogression, assuredly the nineteenth century would have no reason to plume itself in comparison with the sixth B.C. A more earnest attempt to account, on metaphysical grounds, for the more perfect consonance of the smaller ratios was made by the celebrated mathematician, Euler, and his explanation, if such it could be called, long silenced, if it did not satisfy, inquirers. Euler analyses the cause of *pleasure.* We take delight in *order*; it is pleasant to us to observe 'means co-operant to an end.' But, then, the effort to discern order must not be so great as to weary us. If the relations to be disentangled are too complicated, though we may see the order we cannot enjoy it. The simpler the terms in which the order expresses itself, the greater is our delight. Hence, the superiority of the simpler ratios in music over the more complex ones. Consonance, then, according to Euler, was the spiritual pleasure derived from the perception of order without weariness of mind.

But in this theory it was overlooked that Pythagoras

himself, who first experimented on these musical intervals, knew nothing about rates of vibration. It was forgotten that the vast majority of those who take delight in music, and who have the sharpest ears for the detection of a dissonance, are in the condition of Pythagoras, knowing nothing whatever about rates or ratios. And it may also be added that the scientific man who is fully informed upon these points has his pleasure in no way enhanced by his knowledge. Euler's explanation, therefore, does not satisfy the mind, and it was reserved for an eminent German investigator of our own day, after a most profound analysis of the entire question, to assign the physical cause of consonance and dissonance:—a cause which, when once clearly stated, is so simple and so satisfactory as to excite surprise that it remained so long without a discoverer.

Various expressions employed in our previous lectures have already, in part, forestalled Helmholtz's explanation of consonance and dissonance. Let me here repeat an experiment which will, almost of itself, force this explanation upon your attention. Before you are two jets of burning gas, which I can convert into singing flames by enclosing them within two tubes. The tubes are of the same length, and the flames are now singing in unison. By means of this telescopic slider I lengthen slightly one of the tubes; you have now these deliberate beats which succeed each other so slowly that they can with ease be counted. I augment still further the length of the tube. The beats are now more rapid than before: they can barely be counted. And now I would ask the musicians to follow me with attention while I gradually increase their quickness of succession. It is perfectly manifest that the rattle that you now hear differs only in point of rapidity from the slow beats which you heard a moment ago. There is not the slightest breach of the continuity of the beats. We begin slowly, we gradually increase the rapidity, and now the succession is so rapid as to produce that particular grating effect which every musician that hears it would call *dissonance.* I will now reverse the process, and pass from these quick beats to slow ones. I gradually shorten the tube which, a moment ago, was lengthened. The same continuity of the phenomenon is noticed. By degrees these beats separate from each other more and more, until finally they are slow enough to be counted. Thus these singing flames enable us to follow the beats with certainty until they cease to be beats, and are converted into dissonance.

This experiment proves conclusively that dissonance *may* be produced by a rapid succession of beats; and I imagine this cause of dissonance would have been discovered earlier had not men's minds been thrown off the proper track by the theory of 'resultant tones' enunciated by Thomas Young. Young imagined that, when they are quick enough, the beats run together to form a resultant tone. He imagined the linking together of beats to be precisely analogous to the linking together of simple musical impulses, and he was strengthened in this notion by the fact already adverted to, that the first difference tone, that is to say, the loudest resultant tone, corresponded, like the beats, to a rate of vibration equal to the difference of the rates of the two primaries. But the fact is, that the effect of beats upon the ear is altogether different from that of the successive impulses of an ordinary musical tone. In our last lecture I operated with a resultant tone produced by 33 vibrations a second. That tone was perfectly smooth and musical; whereas beats which succeed each other at the rate of 33 per second, are pronounced by the disciplined ear of Helmholtz to be in their condition of most intolerable dissonance. Hence the resultant tone referred to could not be produced by the coalescence of beats. When the beats are slower than 33 they are less disagreeable. They may even become pleasant through imitating the trills of the human voice. With higher rates than 33 the roughness also lessens, but it is still discernible when the beats number 100 a second. Helmholtz fixes the limit at which they totally disappear at 132. Thus we learn that the continuity and smoothness of a tone, produced by ordinary sonorous waves, is perfect at rates of vibration far below that which corresponds to the disappearance of the beats. The impulses of ordinary sonorous waves are, in fact, gently graduated; in the beats, on the contrary, the boundaries of sound and silence are abrupt, and they, therefore, subject the ear to that jerking intermittence which expresses itself to consciousness as dissonance.

Does this theory accord with the facts that have been adduced? Let us in the first place examine our four tuning-forks, to ascertain whether their deportment harmonises with this theory. And here let me remark that we have only to do with the fundamental tones of the forks. Care has been taken that their overtones should not come into play, and they have been sounded so feebly that no resultant tones mingled in any sensible degree with the fundamental tones. Bearing in mind, then, that the beats and the dissonance vanish when the difference of the two rates of vibration is 0; that the dissonance is at its maximum when the beats number 33 per second; that it lessens gradually afterwards, and entirely disappears when the beats amount to 132 per second,—we will analyse the sounds of our forks; beginning with the *octave*.

Here our rates of vibration are,

$$512 - 256; \text{ difference} = 256.$$

It is plain that in this case we can have no beats, the difference being too high to admit of them.

Let us now take the *fifth*. Here the rates of vibration are,

384 − 256 ; difference = 128.

This difference is barely under the number 132 at which the beats vanish: consequently the roughness must be very slight indeed.

Taking the *fourth*, the numbers are,

384 − 312 ; difference = 72.

Here we are clearly within the limit, where the beats vanish, the consequent roughness being quite sensible.

Taking the *major third*, the numbers are,

320 − 256 ; difference = 64.

Here we are still further within the limit, and, accordingly, the roughness is more perceptible.

Thus we see that the deportment of our four tuning-forks is entirely in accordance with the explanation which assigns the dissonance to beats.

And here it is to be remarked that if beats are to be avoided in the case of the octave and of the fifth these intervals must be *pure*: that is to say, for every wave sent forth by the fundamental fork, exactly two must be sent forth by its octave; while, in the case of the fifth, for every two waves sent forth by the one fork, exactly three must be sent forth by the other. If, in the case of the octave, I load either of the forks with a bit of wax I obtain distinct beats. Placing the resonant case of the fundamental fork near my ear, and holding the octave and its case at some distance, the beats are heard; and they are heard as the beats of the fundamental note, the octave apparently making no impression whatever upon the ear. When the octave is pure, the relation of the two notes at any moment continues to be their relation throughout. If, for example, the condensation of one of the shorter waves fall upon the condensation of one of the larger, the condensations will remain associated as long as the ratio is exactly 1 : 2. But if we deviate from this ratio in a small degree, though, at starting, the condensations may coincide, the phases come eventually into opposition, as in the cases so fully illustrated in our last lecture. And, as a consequence of this alternate coincidence and opposition, we must have beats. Similar remarks apply to other intervals than the octave.

We have here confined ourselves to a pair, in each case, of single primary tones; but Helmholtz has shown the important part played by the overtones and resultant tones in the question of musical consonance. In the tuning-forks that we have thus far experimented with, both these classes of tones have been avoided, as I wished to make my first example an easy one. But other sources of musical sounds do not yield these simple tones. The strings of a violin, for example, are rich in overtones, whose interferences must be taken into account when judging of the combination of the sounds of two strings. And be it remarked that the overtones are indispensable to the character of musical sounds. Pure sounds, without overtones, would be like pure water, flat, and dull. The tones, for example, of wide stopped organ-pipes are almost perfectly pure; for the tendency to subdivision is here so feeble, that the overtones of the pipe hardly come into play. But the tones of such pipes, though mellow, would soon weary us; they are without force or character, and would not satisfy the demand of the ear for brightness and energy. In fact, a good musical clang requires the presence of several of the first overtones. So much are these felt to be a necessity that it is usual to associate with the deeper pipes of the organ, shorter pipes which yield the harmonic tones of the deeper one. In this way, where the vibrating body itself is incapable of furnishing the overtones, they are supplied from external sources.

Let us now examine how the action of the overtones agrees with the foregoing theory of consonance and dissonance. We will take, in illustration, the octave which includes the middle A of the piano-forte. This note corresponds to 440 vibrations a second; the C below it corresponds to 264; the C above it to 528. Let us call the former c' and the latter c'', and fixing our attention upon the octave between c' and c'', let us place plainly before ourselves both the fundamental tones and overtones of its various intervals, including all overtones as far as the ninth. First then with regard to the octave, c', c'', its two fundamental tones and their overtones answer respectively to the following rates of vibration :—

		1	:	2	
Fundamental tone		264		528	Fundamental tone
Overtones .	1.	528		1056	
„ .	2.	792		1584	
„ .	3.	1056		2112	
„ .	4.	1320		2640	
„ .	5.	1584		3168	
„ .	6.	1848		3696	
„ .	7.	2112		4224	
„ .	8.	2376		4752	
„ .	9.	2640		5280	

Comparing these tones together in couples, it is impossible to get out of the two series a pair whose difference is less than 264. Hence, as the beats cease to be heard as dissonance when they reach 132, dissonance must be entirely absent from the combination which we have just examined. This octave, therefore, is an absolutely perfect consonance.

Let us now take the interval of a fifth. We have the following fundamental tones and overtones :—

		2	:	3	
Fundamental tone		264		396	Fundamental tone
Overtones .	1.	528		792	
„ .	2.	792		1188	
„ .	3.	1056		1584	
„ .	4.	1320		1980	
„ .	5.	1584		2376	
„ .	6.	1848		2772	
„ .	7.	2112		3168	
„ .	8.	2376		3564	
„ .	9.	2640		3960	

The lowest difference here is 132, which corresponds to the vanishing point of the dissonance. The interval of a fifth in this octave is, therefore, all but perfectly free from dissonance.

Let us now take the interval of a fourth.

		3	:	4	
Fundamental tone		264		352	Fundamental tone
Overtones .	1.	528		704	
,, .	2.	792		1056	
,, .	3.	1056		1408	
,, .	4.	1320		1760	
,, .	5.	1584		2112	
,, .	6.	1848		2464	
,, .	7.	2112		2816	
,, .	8.	2376		3168	
,, .	9.	2640		3520	

Here we have a series of differences each equal to 88, but none lower. This number, though within the vanishing limits of the beats, is still so high as to leave the roughness of the beats very inconsiderable. Still, the interval is clearly inferior to the fifth.

Again, let us take the major third. Here we have

		4	:	5	
Fundamental tone		264		330	Fundamental tone
Overtones .	1.	528		660	
,, .	2.	792		990	
,, .	3.	1056		1320	
,, .	4.	1320		1650	
,, .	5.	1584		1980	
,, .	6.	1848		2310	
,, .	7.	2112		2640	
,, .	8.	2376		2970	
,, .	9.	2640		3300	

There are here several differences, each equal to 66. The beats are nearer the maximum dissonance than in the last case, and the interval, therefore, is less perfect as a consonance than the last.

We will now try the minor third. Here we have

		5 :	6	
Fundamental tone		264	316·8	Fundamental tone
Overtones .	1.	528	433·6	
" .	2.	792	950·4	
" .	3.	1056	1267·2	
" .	4.	1320	1584·0	
" .	5.	1584	1900·8	
" .	6.	1848	2217·6	
" .	7.	2112	2534·4	
" .	8.	2376	2851·2	
" .	9.	2640	3168·0	

Between several pairs of these tones we have a difference of 53 vibrations. This difference implies a greater disturbance by beats in the case of this interval than in the case of the fifth, or of the fourth, or of the major third. Hence, the minor third is inferior as a consonance to all those intervals.

Thus do we find that as the numbers expressing the ratio of the vibrations become larger, the disturbing influence of the beats enters more and more into the interval. The result, it is manifest, entirely harmonises with the explanation that refers dissonance to beats.

Helmholtz has attempted to represent this result graphically, and from his work I copy the two adjacent diagrams. He assumes, as already stated, the maximum dissonance to correspond to 33 beats per second; and he seeks to express different degrees of dissonance by lines of different lengths. The horizontal line *c′ c″*, fig. 152, represents the range of the musical scale that we have just been analysing, and the distance from any point of this line to the curve above it represents the dissonance corresponding to that point. The pitch here is supposed to ascend continuously, and not by jumps. Supposing, for example, two performers on the violin to start with the same note *c′*, and that, while one of them continues to sound that note, the other gradually and continuously shortens his string, thus gradually raising its pitch up to c''. The effect upon the ear would be repre-

FIG. 152.

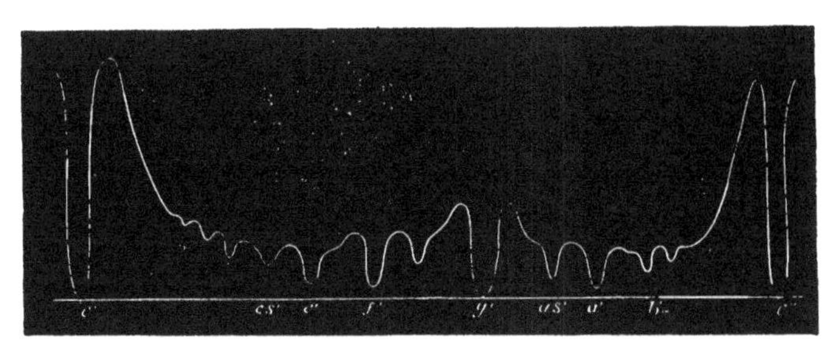

sented by the irregular curved line in fig. 152. Soon after the unison, which is represented by contact at c', is departed from, the curve suddenly rises, showing the dissonance here to be the sharpest of all. At e', the curve approaches the straight line c' c'', and this point of the curve corresponds to the major third. At f' the approach is still nearer, and this point corresponds to the fourth. At g' the curve almost touches the straight line, which indicates that at this point, which corresponds to the fifth, the dissonance almost vanishes. At a' we have the major sixth; while at c'', where the one note is an octave above the other, the dissonance also vanishes. The *e s'* and the *a s'* of this diagram are the German names of a flat third and a flat sixth.

When the fundamental note is c'', that is, the higher octave of the c' which formed our last starting-point, the various degrees of consonance and dissonance are those shown in fig. 153. Beginning with the unison c''—c'', and gradually elevating the pitch of one of our strings, till it reaches c''', the octave of c'', the curved line represents the effect upon the ear. We see, from both these curves, that dissonance is the general rule, and that only at certain definite points do we find the dissonance vanish, or become so decidedly enfeebled as not to destroy the harmony. These points correspond to the places where the numbers ex-

FIG. 153.

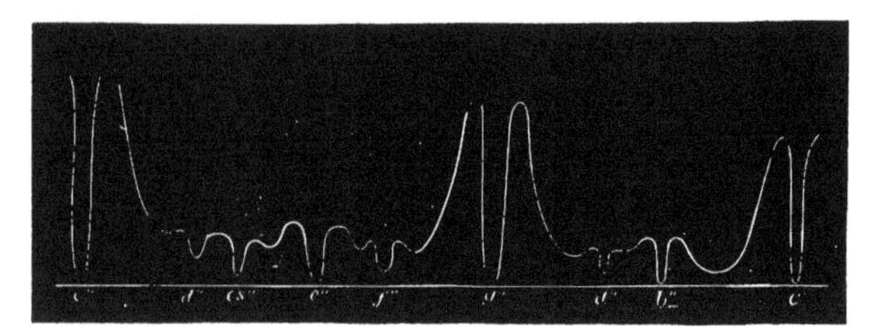

pressing the ratio of two rates of vibration are small whole numbers. It must be remembered that these curves are constructed on the supposition, that the beats are the cause of the dissonance; and the agreement between calculation and experience sufficiently demonstrates the truth of the assumption.

You have thus accompanied me to the verge of the Physical portion of the science of Acoustics, and into the musical portion it is not my vocation to lead you. I will only add, that in comparing three or more sounds together, that is to say, in choosing them for *chords*, we are guided by the principles just mentioned. We choose sounds which are in harmony with the fundamental sound and with each other In choosing a series of sounds for combination two by two, the simplicity alone of the ratios would lead us to fix on those expressed by the numbers 1, $\frac{5}{4}$, $\frac{4}{3}$, $\frac{3}{2}$, $\frac{5}{3}$, 2; these being the simplest ratios that we can have within an octave. But when the notes represented by these ratios are sounded in succession, it is found that the intervals between 1 and $\frac{5}{4}$, and between $\frac{5}{3}$ and 2 are wider than the others, and require the interpolation of a note in each case. The notes chosen are such as form chords, not with the fundamental tone, but with the note $\frac{3}{2}$ regarded as a fundamental tone. The ratios of these two notes with the fundamental are $\frac{9}{8}$ and $\frac{15}{8}$. Interpolating these, we have the eight notes of the natural or diatonic scale expressed by the following names and ratios:—

Names.	C.	D.	E.	F.	G.	A.	B.	C'.
Intervals.	1st.	2nd.	3rd.	4th.	5th.	6th.	7th.	8th.
Rates of vibration.	1,	$\frac{9}{8}$,	$\frac{5}{4}$,	$\frac{4}{3}$,	$\frac{3}{2}$,	$\frac{5}{3}$,	$\frac{15}{8}$,	2.

Multiplying these ratios by 24 to avoid fractions, we obtain the following series of whole numbers, which express the relative rates of vibration of the notes of the diatonic scale.

24, 27, 30, 32, 36, 40, 45, 48.

The meaning of the terms third, fourth, fifth, &c., which we have already so often applied to the musical intervals, is now apparent; the term has reference to the position of the note in the scale.

In our second lecture I referred to, and in part illustrated, a method devised by M. Lissajous for studying musical vibrations. By means of a beam of light reflected from a mirror attached to a tuning-fork, the fork was made to write the story of its own action. In our last lecture the same method was employed to illustrate optically the phenomena of beats. I now propose to apply it to the study of the composition of the vibrations which constitute the principal intervals of the diatonic scale. We must, however, prepare ourselves for the thorough comprehension of this subject by a brief preliminary examination of the vibrations of a common pendulum.

Such a pendulum hangs before you. It consists of a wire carefully fastened to a plate of iron at the roof of the house, and bearing a copper ball weighing 10 lbs. I draw the pendulum thus aside and let it go; it oscillates to and fro almost in the same plane.

I say 'almost,' because it is practically impossible to

suspend a pendulum without some little departure from perfect symmetry around its point of attachment. In consequence of this, the weight deviates sooner or later from a straight line, and describes an oval more or less elongated. This circumstance presented a serious difficulty some years ago to those who wished to repeat M. Foucault's celebrated experiment, demonstrating the rotation of the earth.

Nevertheless, in the case now before us, the pendulum is so carefully suspended, that its deviation from a straight line is not at first perceptible. Let us suppose the amplitude of its oscillation to be represented by the dotted line *a b*, fig. 154. The point *d*, midway between *a* and *b*, is the pendulum's point of rest. When drawn aside from this point to *b*, and let go, it will return to *d*, and in virtue of its momentum will pass on to *a*. There it comes to momentary rest, and returns through *d* to *b*. And thus it will continue to oscillate until its motion is expended.

Fig. 154.

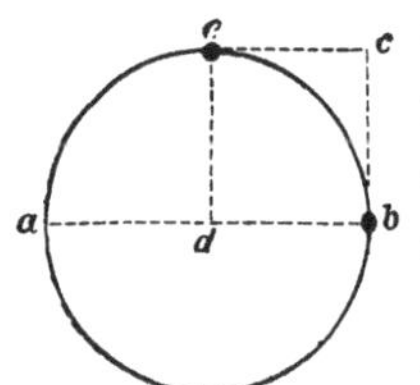

The pendulum having first reached the limit of its swing at *b*, let us suppose a push in a direction perpendicular to *a b* imparted to it; that is to say, in the direction *b c*. Supposing the time required by the pendulum to swing from *b* to *a* to be one second,* then the time required to swing from *b* to *d* will be half a second. Suppose, further, the force applied at *b* to be such as would carry the bob, if free to move in that direction alone, to *c* in half a second, and that the distance *b c* is equal to *b d*, the question then occurs where will the bob really find itself at the end of half a second? It is perfectly manifest that both

* This supposition is of course made for the sake of simplicity, the real period of oscillation of a pendulum 28 feet long being between two and three seconds.

forces are satisfied by the pendulum reaching the point *e*, exactly opposite the centre *d*, in half a second. To reach this point it can be shown that it must describe the circular arc *b e*, and it will pursue its way along the continuation of the same arc, to *a*, and then pass round to *b*. Thus, by the rectangular impulse the oscillation is converted into a rotation, the pendulum describing a circle, as shown in fig. 155.

FIG. 155.

If the force applied at *b* be sufficient to urge the bob in half a second through a greater distance than *b c*, the pendulum will describe an ellipse, with the line *a b* for its smaller axis; if, on the contrary, the force applied at *b* urge the pendulum in half a second through a distance less than *b c*, the bob will describe an ellipse, with the line *a b* for its greater axis.

Let us now enquire what occurs when the rectangular impulse is applied at the moment the ball is passing through its position of rest at *d*.

Supposing the pendulum to be moving from *a* to *b*, fig. 156, and that at *d* a shock is imparted to it sufficient of itself to carry it in half a second to *c*; it is here manifest that the resultant motion will be along the straight line *d g* lying between *d b* and *d c*. The pendulum will return along this line to *d*, and pass on to *h*. In this case, therefore, the pendulum will describe a straight line, *g h*, oblique to its original direction of oscillation.

Supposing the direction of motion at the moment the push is applied to be from *b* to *a*, instead of from *a* to *b*, it is manifest that the resultant here will also be a straight line oblique to the primitive direction of oscillation, but its obliquity will be that shown in fig. 157.

When the impulse is imparted to the pendulum neither at the centre nor at the limit of its swing, but at

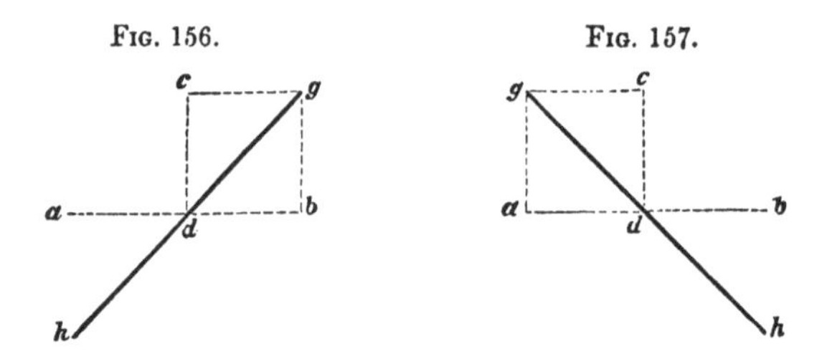

FIG. 156. FIG. 157.

some point between both, we obtain neither a circle nor a straight line, but something between both. We have, in fact, a more or less elongated ellipse with its axis oblique to $a\ b$, the original direction of vibration. If, for example, the impulse be imparted at d', fig. 158, while the pendulum is

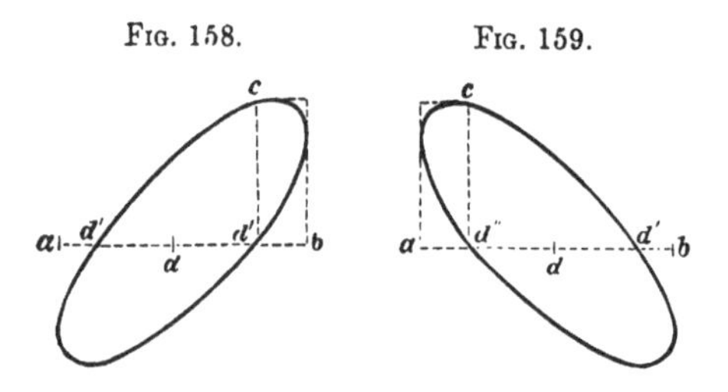

FIG. 158. FIG. 159.

moving *towards* b, the position of the ellipse will be that shown in fig. 158; but if the push at d' be given when the motion is towards a, then the position of the ellipse will be that represented in fig. 159.

In the case of the rod employed in our fourth lecture to illustrate Mr. Wheatstone's Kaleidophone, we had a combination of vibrations similar to that here exhibited by the pendulum, and to this combination were due the figures of the circle, the ellipse, and the straight line obtained when the rod was caused to oscillate. By the method of M. Lissajous we can combine the rectangular vibrations of two tuning-forks,—a subject which I now wish to illustrate before you. In front of this electric lamp, L, fig. 160, is placed a large tuning-fork, T′, provided with a mirror, on which a narrow beam of light, L T′, is permitted to fall. The beam is thrown back, by reflection, upon the audience. The tuning-fork, T′, you observe, is

FIG. 160.

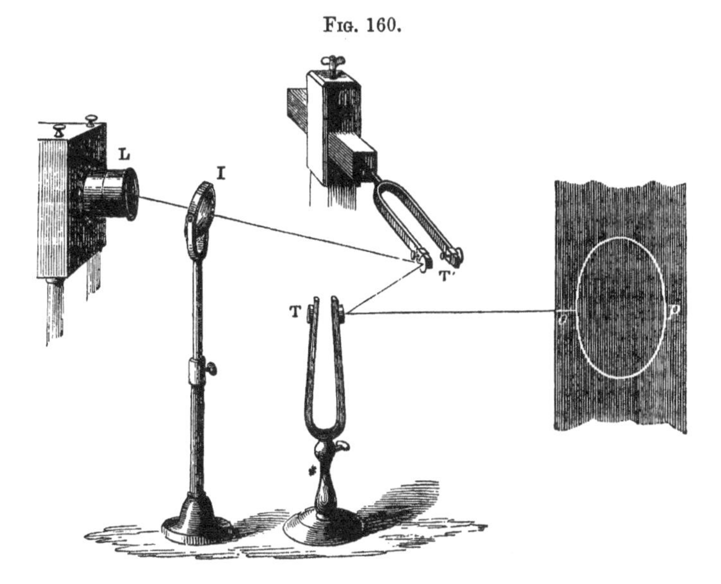

fixed in its stand (only partially shown in the figure) in a horizontal position, and will, therefore, when it vibrates, tilt the beam to and fro laterally. In the path of the reflected beam I place a second tuning-fork, T, also furnished with a mirror. The second fork, however, is upright, and when it vibrates it tilts, as you know, the luminous beam up and down. At the present moment both forks are motionless, the beam of light is reflected from the mirror of the horizontal to that of the vertical fork, and from the latter it is thrown back upon the screen, on which it prints a brilliant disc. I now agitate the upright fork, leaving the other motionless. The disc is drawn out into this fine luminous band, 3 feet long. I now sound the second fork, and the straight band is instantly transformed into this splendid ring, *o p*, fig. 160, 36 inches in diameter. What have we done here? Exactly what we did in our first experiment with the pendulum. We have caused a beam of light to vibrate simultaneously in two directions, and have accidentally hit upon the combination when one fork has just reached the limit of its swing, and come momentarily to rest, while the other is passing with its maximum velocity through its position of equilibrium.

That the *circle* was obtained is, as I have said, a mere accident; but it was a fortunate accident, as it enables us to see the exact similarity between the motion of the beam and the motion of the pendulum. I stop both forks and agitate them afresh. You have now an ellipse before you with its axis oblique. After a few trials we obtain the straight line, indicating that both the forks then pass simultaneously through their positions of equilibrium. In this way, by combining the vibrations of the two forks, we produce all the figures which we obtained with the pendulum.

When the vibrations of the two forks are, in all respects, absolutely alike, whatever the figure may be which is first traced upon the screen, it remains unchanged in form, diminishing only in size as the motion is expended. But the slightest difference in the rates of vibration destroys this fixity of the image. I endeavoured before the lecture to render the unison between these two forks as perfect as possible, and hence you have observed very little alteration in the shape of the figure. But by moving a little weight along the prong of either fork, or by attaching to either of them a bit of wax, I impair the unison, and now you have the figure slowly passing from a straight line into an oblique ellipse, thence into a circle; after which it narrows again to an ellipse with an opposed obliquity, then passes into a straight line, the direction of which is at right angles to the first direction: finally it passes, in the reverse order, through the same series of figures to the straight line with which we commenced. The interval between two successive identical figures is the time in which one of the forks succeeds in beating the other by one complete vibration. Loading the fork still more heavily, we have more rapid changes; the straight line, ellipse, and circle being passed through in quick succession. At times the luminous curve exhibits an apparent stereoscopic depth which renders it difficult to believe that we are not looking at a solid ring of white-hot metal.

By causing the mirror of the fork, T, to rotate through

FIG. 161.

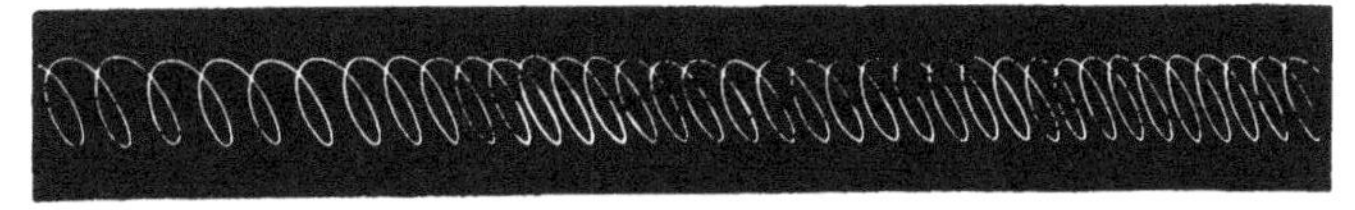

a small arc, I draw the steady circle out into this splendid luminous scroll which stretches right across the screen, fig. 161. The same experiment made with the changing figure gives us a scroll of irregular amplitude, fig. 162.*

FIG. 162.

We have next to combine the vibrations of two forks, one of which oscillates with twice the rapidity of the other, that is, to determine the figure corresponding to the combination of a note and its octave. To prepare

* This figure corresponds to the interval 15: 16. For it and some other figures, I am indebted to that excellent mechanician, M. König of Paris.

ourselves for the mechanics of the problem, we must resort once more to our pendulum; for it also can be caused to oscillate in one direction twice as rapidly as in another. By a complicated mechanical arrangement this might be done in a very perfect manner, but at present I prefer simplicity to completeness. I have therefore permitted

FIG. 163.

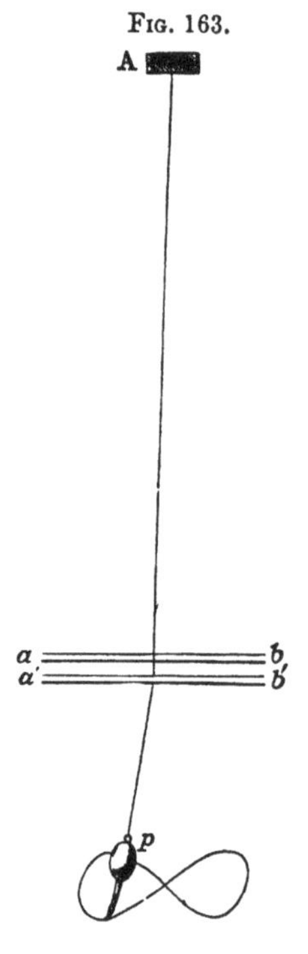

the wire of our pendulum to descend from its point of suspension, A, fig. 163, midway between two horizontal glass rods, *a b*, *a′ b′*, supported firmly at their ends, and about an inch asunder. The rods cross the wire at a height of 7 feet above the bob of the pendulum. The whole length of the pendulum being 28 feet, the glass rods intercept one-fourth of this length. I draw the pendulum aside in the direction of the rods, *a b*, *a′ b′*, and let it go. It oscillates freely between them, its whole length swinging to and fro. I bring it to rest and draw it aside in a direction perpendicular to the last; a length of 7 feet only can now oscillate, and by the laws of oscillation a pendulum 7 feet long vibrates with twice the rapidity of a pendulum 28 feet long. In order to show you the figure described by the combination of these two rates of vibration, I have attached to the copper ball, *p*, a camel's hair pencil, fig. 163, which rubs lightly upon a glass plate placed on black paper. Over the plate I strew white sand, and allowing the pendulum to oscillate as a whole, the sand is rubbed away along a straight line which represents the amplitude of the vibration. Let *a b*, fig. 164, represent this line, which, as before, we assume to be described in one second. When the pendulum is at the limit, *b*, of its swing, let a rectangular impulse be imparted to it sufficient to carry it to *c* in one-fourth of a second. If this were the only impulse acting on the pendulum, the bob would reach *c* and return to *b* in half a second. But under the actual circumstances it is also urged towards *d*, which point, through the vibration of the whole pendulum, it ought also to reach in half a second. Both vibrations, therefore, require that the bob shall reach *d* at the same moment, and to do this it will have to describe the curve *d c′ b*. Again, in the time required by the long pendulum to pass from *d* to *a*, the short pendulum will pass *to and fro* over the half of its excursion; both vibrations must therefore reach *a* at the same moment, and to accomplish this the pendulum describes the lower curve between *d* and *a*. It is manifest that these two curves will repeat themselves at the opposite sides of *a b*, the combination of both vibrations producing finally this figure of 8, which you now see fairly drawn upon the sand before you.

FIG. 164.

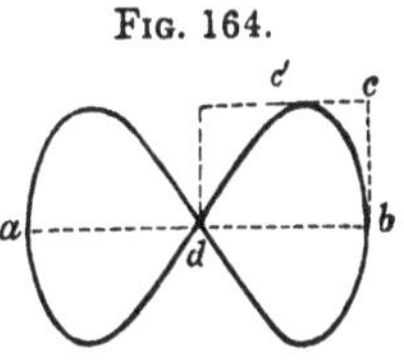

The same figure is obtained if the rectangular impulse be imparted when the pendulum is passing its position of rest, *d*. I have here supposed the time occupied by the pendulum in describing the line *a b* to be one second. Let us suppose three-fourths of the second exhausted, and the pendulum at *d′*, fig. 165, in its excursion towards *b*; let the rectangular impulse then be imparted to it, sufficient to carry it to *c* in one-fourth of a second. Now the long pendulum requires that it should move from *d′* to *b* in one-fourth of a second; both impulses

FIG. 165.

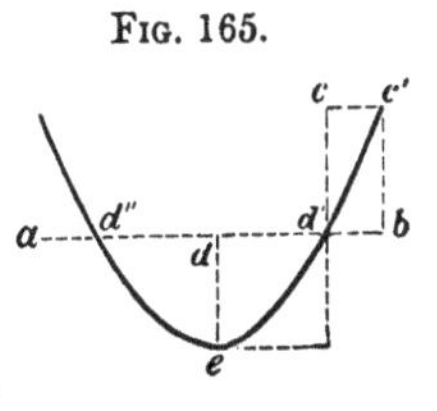

are therefore satisfied by the pendulum taking up the position c' at the end of a quarter of a second. To reach this position it must describe the curve $d' c'$. It will manifestly return along the same curve, and at the end of another quarter of a second find itself again at d'. From d' to d the long pendulum requires a quarter of a second. But at the end of this time the short pendulum must be at the lower limit of its swing: both requirements are satisfied by the pendulum being at e. We thus obtain one arm, $c' e$, of a curve which repeats itself to the left of e, so that the entire curve due to the combination of the two vibrations is that represented in fig. 165. This figure is called by geometers a parabola; whereas the figure of 8 before obtained is called a lemniscata.

We have here supposed that at the moment when the rectangular impulse was applied the motion of the pendulum was *towards* b: if it were towards a, we should obtain the parabola inverted as shown in fig. 166.

FIG. 166.

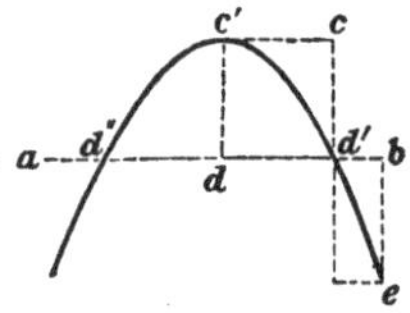

Supposing, finally, the impulse to be applied, not when the pendulum is passing through its position of equilibrium, not when it is passing a point corresponding to three-fourths or one-fourth of the time of its excursion, but at some other point in the line, $a\ b$, between its end and centre. Under these circumstances we should have neither the parabola, nor the perfectly symmetrical figure of 8, but a distorted 8, which would assume positions depending on the direction of the motion at the moment when the rectangular impulse is imparted.

And now we are prepared to witness with profit the combined vibrations of our two tuning-forks, one of which sounds the octave of the other. Permitting the vertical fork, T, fig. 160, to remain undisturbed in front of the lamp, I oppose to it a horizontal fork, which vibrates with twice the rapidity. The first passage of the bow across the two forks reveals to you the exact similarity of this combination, and that of our pendulum. A very perfect figure of 8 is now described upon the screen. Before the lecture I fixed the vibrations of these two forks as nearly as possible to the ratio of 1 : 2, and the steadiness of the figure upon the screen indicates the perfection of the tuning. I stop both forks, and again agitate them. Now you have the distorted 8 upon the screen. Again I stop the forks, and after a few trials bring out the parabola. In all these cases the figure remains fixed upon the screen; but I now attach a morsel of wax to one of the forks. The figure is steady no longer, but passes from the perfect 8 into the distorted one, thence into the parabola, from which it afterwards opens out to an 8 once more. By augmenting the discord, we can render those changes as rapid as we please.

When the 8 is steady on the screen a rotation of the mirror of the fork, T, produces on the screen the scroll shown in fig. 167.

FIG. 167.

I will now combine two forks vibrating in the ratio of 2 : 3. Observe in the first instance the admirable steadiness of the figure produced by the compounding of these two rates of vibration. I now attach with wax to one of the forks a fourpenny piece, the steadiness ceases, and we have this apparent rocking to and fro of the luminous figure. Passing on to intervals of 3 : 4, 4 : 5, and 5 : 6, the figures become more intricate as we proceed. The last combination, 5 : 6, is so entangled, that to see the figure plainly a very narrow band of light must be employed. The great distance between the forks and the screen also helps us to unravel the complication. And here it is worth noting that *when the figure is fully developed* the loops along the vertical and horizontal edges express the ratio of the combined vibrations. In the octave, for example, we have two loops in one direction and one in another; in the fifth two loops in one direction, and three in another. When the combination is as 1 : 3 the luminous loops are also as 1 : 3. The changes which some of these figures undergo when the tuning is not perfect, or when one fork is purposely loaded, are extremely remarkable. In this last case of 1 : 3, for example, it is difficult at times not to believe that you are looking at a solid link of white-hot metal. The figure exhibits a depth apparently incompatible with its being traced upon a plane surface.

Before you, fig. 169, is suspended a diagram of these beautiful figures, including combinations from 1 : 1 to 5 : 6. In each case the characteristic phases of the vibration are shown, through all of which each figure passes when the interval between the two forks is not pure: I also add here, fig 168, two appearances of the combination 8 : 9.

Fig. 168.

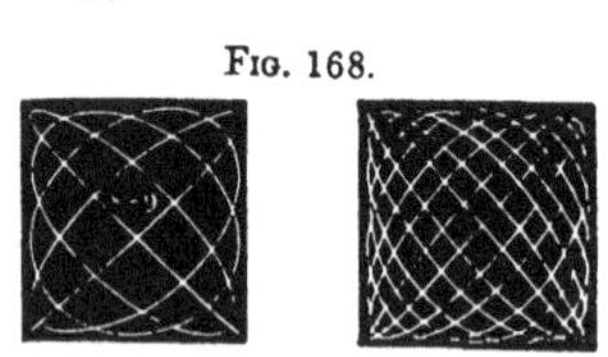

All these figures may be obtained from the vibrations of the selfsame body. I have here a rod which, instead of being round, is rectangular, being so shaped as to vibrate in one direction twice as quickly as in the other. It is furnished with a silvered bead which can be illuminated, and the image of which can be thrown upon our screen.

Fig. 169.

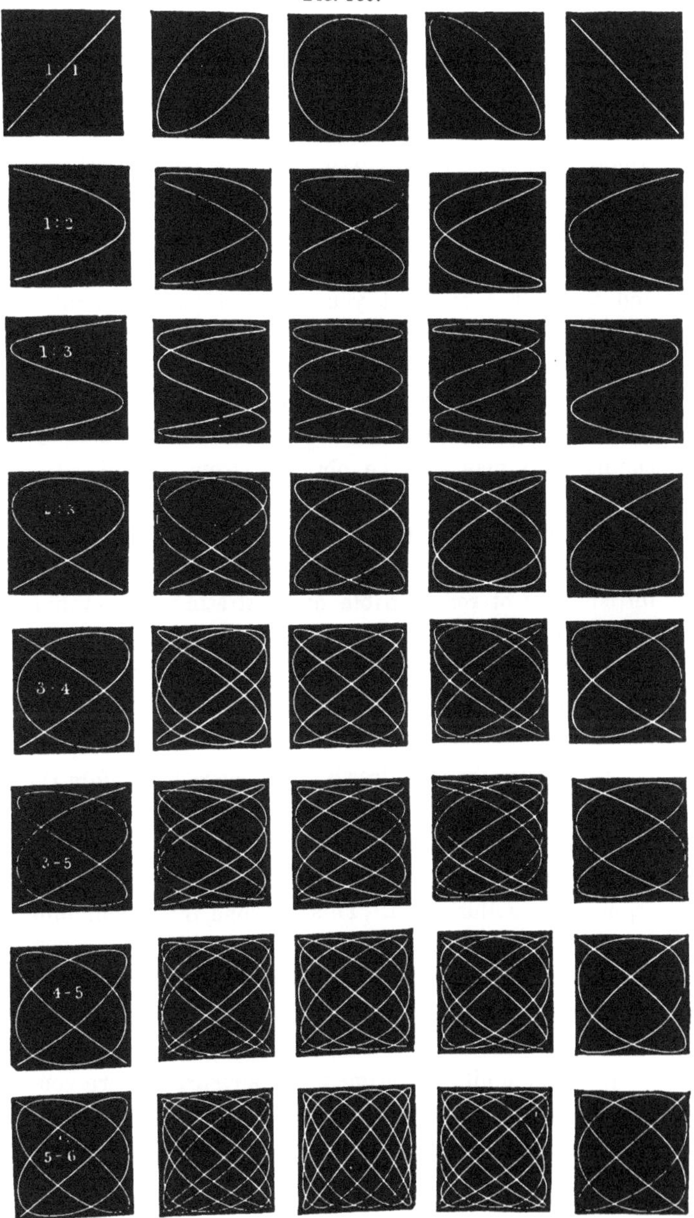

The rod is now fixed in a vice; a lens is in front of it, and you see the spot of light upon the screen. I draw the rod aside and suddenly liberate it; you observe the figures it describes. Were the rates of vibration in both directions accurately in the ratio of 1 : 2, the figure would remain perfectly fixed; but, inasmuch as this relation is only approximately secured, you see the figure first as a distorted 8, then as a perfect 8, then again as an 8 distorted in a different direction from its first, and now for a moment you see it a perfect parabola. Here we have precisely the same succession of figures as were obtained with our two tuning-forks vibrating in the ratio of 1 : 2, and thrown slightly out of tune. All the other figures produced by the combination of two tuning-forks may also be obtained from the vibrations of single rods, so formed that they shall oscillate with different rapidities in two rectangular directions.

Mr. Wheatstone has perfected his Kaleidophone by the introduction of rods capable of illustrating the combinations of all these musical intervals. He has also devised various other extremely ingenious methods for the compounding of vibrations.

Our labours are now drawing to a close, but before they terminate, and to make them terminate properly, I would ask you to call to mind the experiments of our third lecture, by which the division of a string into its harmonic segments was illustrated. This was done, if you remember, by means of little paper riders placed across the string, which were unhorsed, or not, according as they occupied a ventral segment or a node. I now wish to illustrate, further than I have hitherto done, a point of some importance, by another and similar experiment of Sauveur. Before you is the sonometer, which was described in our third lecture. Along it, instead of one, I have stretched two strings, about three inches asunder. By means of a key I alter the tension of the strings, continually sounding both of them until they are brought into perfect unison. And now I place a little paper rider upon the middle of one of them, and agitate the other. What occurs? The vibrations of the sounding string are communicated to the bridges on which it rests, and through the bridges to the other string. The individual impulses are very feeble, but, because the two strings are in unison, the impulses can so accumulate as finally to toss the rider off the untouched string.

Every experiment with the riders and a single string executed in our third lecture, may be repeated with there two unisonant strings. Let us, for example, damp one of the strings at a point one fourth of its length from one of its ends; and let us place the red and blue riders formerly employed, not on the nodes and ventral segments of the damped string, but at points upon the other exactly opposite to those nodes and ventral segments. When the bow is passed across the shorter segment of the damped string, the five red riders on the adjacent string are unhorsed, while the four blue ones remain tranquilly in their places. By relaxing one of the strings, I throw it out of unison with the other. All my efforts to unhorse the riders are now unavailing. That accumulation of impulses which unison alone renders possible cannot here take place, and the consequence is, that however great the agitation of the one string may be, it fails to produce any sensible effect upon the other.

The influence of synchronism may be illustrated in a still more striking manner by means of two tuning-forks which sound the same note. I place two such forks mounted on their resonant supports upon the table, 18 inches asunder, and draw the bow vigorously across one of them. The other fork has remained untouched. I now stop the agitated fork; the sound is enfeebled but by no means quenched. Through the air and through the wood the vibrations have been conveyed from fork to fork, and the untouched fork is the one which you now hear. I attach a morsel of wax to one of the forks, and sound it again; its power of influencing the other is gone; the change in the rate of vibration, small as it is, has destroyed the sympathy between the two forks, and no response is now possible. I remove the wax, and the untouched fork responds as before. This communication of vibrations through the wood and air may be obtained when the forks, mounted on their cases, stand several feet apart. But the vibrations may also be communicated through the air alone. I stop one of the forks, and cause the other to vibrate vigorously. Holding the case of the vibrating fork in my hand, I bring one of its prongs near the unvibrating one, placing the prongs back to back, but allowing a space of air to exist between them. Light as is the vehicle, the accumulation of impulses secured by the perfect unison of the two forks enables the one to set the other in vibration. I suddenly extinguish the sound of the agitated fork, but the fork which a moment ago was silent, continues sounding, having taken up the vibrations of its neighbour.

Removing one of the forks from its resonant case, I strike it against a pad, and throw it into strong vibration. Held thus free in the air, its sound is inaudible. But I now bring it close to the silent mounted fork. Out of the silence rises this full mellow sound, which is due, not to the fork originally agitated, but to its sympathetic neighbour.

Various other examples of the influence of synchronism which have been brought forward in these lectures, will occur to you here; and cases of the kind might be readily multiplied. If two clocks, for example, with pendulums of the same period of vibration, be placed against the same wall, and if one of the clocks be set going and the other not, the ticks of the moving clock, transmitted through the wall, will start its neighbour. The pendulum, moved by a single tick, swings through an extremely minute arc, but it returns to the limit of its swing just in time to receive another impulse. By the continuance of this process, the impulses so add themselves together as finally to set the clock a-going. It is by this timing of impulses that a properly pitched voice can cause a glass to ring, and that the sound of an organ can break a particular window-pane.

I dwell upon this subject here for the purpose of rendering the manner in which sonorous motion is communicated to the auditory nerve more intelligible to you That nerve is in all probability set in motion by bodies associated with it, which are capable of entering into sympathetic vibration with the different waves of sound. In the organ of hearing in man we have first of all the external orifice of the ear, which is closed at the bottom by the circular tympanic membrane. Behind that membrane is the cavity called the drum of the ear, this cavity being separated from the space between it and the brain by a bony partition, in which there are two orifices, the one round and the other oval. These orifices are also closed by fine membranes. Across the cavity of the drum stretches a series of four little bones: the first, called the *hammer*, is attached to the tympanic membrane; the second, called the *anvil*, is connected by a joint with the hammer; a third little round bone connects the anvil with the *stirrup bone*, which has its oval base planted against the membrane of the oval orifice above referred to. The base of the stirrup bone abuts against this membrane, almost covering it, and leaving but a narrow rim of the membrane surrounding the bone. Behind the bony partition, and between it and the brain, we have the extraordinary organ called the *labyrinth*, which is filled with water, and over the lining membrane of which the terminal fibres of the auditory nerve are distributed. When the tympanic membrane receives a shock, that shock is transmitted through the series of bones above referred to, and is concentrated on the membrane against which the base of the stirrup bone is planted. That membrane transfers the shock to the water of the labyrinth, which, in its turn, transfers it to the nerves.

The transmission, however, is not direct. At a certain place within the labyrinth exceedingly fine elastic bristles, terminating in sharp points, grow up between the terminal nerve fibres. These bristles, discovered by Max Schultze, are eminently calculated to sympathise with those vibrations of the water which correspond to their proper periods. Thrown thus into vibration, the bristles stir the nerve fibres which lie between their roots, and excite audition. At another place in the labyrinth we have little crystalline particles called *otolithes*—the Hörsteine of the Germans—embedded among the nervous filaments, and which, when they vibrate, exert an intermittent pressure upon the adjacent nerve fibres, thus exciting audition. The otolithes probably subserve a different purpose from that fulfilled by the bristles of Schultze. They are fitted, by their weight, to accept and prolong the vibrations of evanescent sounds, which might otherwise escape attention. The bristles of Schultze, on the contrary, because of their extreme lightness, would instantly yield up an evanescent motion, while they are eminently fitted for the transmission of continuous vibrations. Finally, there is in the labyrinth a wonderful organ, discovered by the Marchese Corti, which is to all appearance a musical instrument, with its chords so stretched as to accept vibrations of different periods, and transmit them to the nerve filaments which traverse the organ. Within the ears of men, and without their knowledge or contrivance, this lute of 3,000 strings* has existed for ages, accepting the music of the outer world, and rendering it fit for reception by the brain. Each musical tremor which falls upon this organ selects from its tensioned fibres the one appropriate to its own pitch, and throws that fibre into unisonant vibration. And thus, no matter how complicated the motion of the external air may be, those microscopic strings can analyse it and reveal the constituents of which it is composed. In these concluding remarks I have endeavoured to place before you, in a few words, the views now entertained by the most eminent authorities regarding the transmission of sonorous motion to the auditory nerve. I do not ask you to consider these views as established, but only as probable. They present the phenomena in a connected and intelligible form, and should they be doomed to displacement by a more correct or comprehensive theory, it will assuredly be found that the wonder is not diminished by the substitution of the truth.

* According to Kölliker, this is the number of fibres in Corti's organ.

INDEX.

LONDON

PRINTED BY SPOTTISWOODE AND CO.

NEW-STREET SQUARE

www.ingramcontent.com/pod-product-compliance
Ingram Content Group UK Ltd.
Pitfield, Milton Keynes, MK11 3LW, UK
UKHW040658180125
453697UK00010B/259